Y0-ABI-675

THE ROLE OF NITROGEN
IN
GRASSLAND PRODUCTIVITY

The Role of Nitrogen in Grassland Productivity

A review of information from temperate regions

by

D. C. WHITEHEAD

The Grassland Research Institute, Hurley

BULLETIN 48

Commonwealth Bureau of Pastures and Field Crops
Hurley, Berkshire, England

Commonwealth Agricultural Bureaux

First published in 1970
by the
Commonwealth Agricultural Bureaux,
Farnham Royal, Bucks, England
Price £2 5s. (£2.25) post free

This and other publications of the
Commonwealth Agricultural Bureaux
can be obtained through any major bookseller
or direct from :
Commonwealth Agricultural Bureaux
Central Sales, Farnham Royal, Bucks, England

SBN 85198 015 5

Set in Baskerville and printed in Great Britain by
The Cambrian News (Aberystwyth) Ltd,
Aberystwyth

CONTENTS

PART III

EFFECTS OF FERTILIZER NITROGEN ON THE COMPOSITION AND QUALITY OF HERBAGE

FOREWORD

It is generally considered that in many countries, including Britain, there is a greater potential for increased agricultural production from grassland than from arable land. However, a rational approach to realizing this potential must be based on an understanding of the factors that limit grassland productivity. Among these factors, the key role of nitrogen is now well recognized and has been the subject of much research the world over. Studies of the nitrogen relations of grassland have occupied an important place in the research programme of the Grassland Research Institute and a considerable amount of data has been built up over the past 15 years. It was my concern to see these results summarized and set in the context of progress and developments elsewhere that gave rise to my proposal that Dr. Whitehead should prepare a review of the subject and led, in collaboration with the Commonwealth Bureau of Pastures and Field Crops, to the publication of this book. The previous lack of a published comprehensive account of the soil, plant, animal and management aspects of the nitrogen economy of grassland suggests that this review will be of considerable and widespread value to research workers and to those giving advice on practical aspects of grassland management. The background information provided has already been taken into account in planning the priorities for future research here, and has contributed to my opinion that further advances in our knowledge of the nitrogen economy of grassland will be best achieved by two types of experimentation—firstly, field investigations involving a wide range of treatments and a comprehensive monitoring of soil and climatic conditions and, secondly, detailed chemical investigations of individual nitrogen transformations.

E. K. WOODFORD.
Director, Grassland Research Institute, Hurley.

May, 1970.

Acknowledgements

I should like to acknowledge the encouragement given by Professor E. K. Woodford, Director of the Grassland Research Institute, Hurley, who originally proposed that this review should be prepared.

I am indebted to a number of colleagues for valuable suggestions and advice. In particular, I would like to thank Dr. J. K. R. Gasser of Rothamsted Experimental Station, Dr. L. H. P. Jones of the Grassland Research Institute, Hurley, and Emeritus Professor G. W. Leeper of the University of Melbourne for their constructive comments on the typescript ; Mr. D. W. Cowling, Mr. T. E. Williams and Dr. C. R. Clement of the Grassland Research Institute for helpful discussions ; Mr. C. L. Skidmore and Mr. P. J. Boyle of the Commonwealth Bureau of Pastures and Field Crops for their continued interest and for editing the text, and Mr. C. G. Waldock, O.B.E., of the Bureau, for his assistance with the bibliography and for preparing the subject index.

Permission to quote unpublished data has been kindly given by Mr. C. H. Mudd, Director of Great House Experimental Husbandry Farm, Lancashire, by Mr. P. G. Shaw, Chief Grassland Officer, North Wyke Experimental Station, Devon, by Dr. J. T. Braunholtz, Director of Jealott's Hill Research Station, Berkshire, and by colleagues at the Grassland Research Institute, Hurley. Dr. J. S. Brockman and Mr. A. J. Low have kindly allowed me to see copies of papers in advance of publication.

INTRODUCTION

In all parts of the world, the supply of nitrogen (N) is of paramount importance for the productivity of grassland. In temperate regions, herbage yield increases are generally proportional to the supply of fertilizer N up to rates as high as 300–450 lb N/acre per year. The magnitude of the yield increase is influenced, however, by a large number of factors, some susceptible to management control and others not. Where high rates of fertilizer N are applied, there may be appreciable loss of N from the sward to the atmosphere or through leaching. In addition to increasing the yield of herbage, the application of fertilizer N is liable to modify its chemical composition and hence its quality for livestock nutrition.

Since grassland is so responsive to fertilizer N, many investigations have been carried out into the various aspects of the N economy of grass swards and the effects of fertilizer N on herbage yield and composition. It is the aim of this publication to bring together in summarized form the most important results of these investigations and to relate them to one of the three themes reviewed in the three main sections of the book. Part I reviews the transformations of N that occur in grassland ecosystems, Part II the yield response of grass swards to the application of fertilizer N as influenced by various environmental and management factors, and Part III the effects of fertilizer N on the composition and quality of herbage.

This review is confined to temperate regions; aspects of the N economy of tropical regions have been reviewed in an earlier book (Bulletin 46) in this series. Much of the work relevant to this review, particularly that concerned with the response of grass to fertilizer N, has in fact been carried out in the cool-temperate regions of north-western Europe, and while the results obtained may be expected to apply to other areas of similar climate, they may not apply without qualification to warm-temperate regions.

In preparing this review, an attempt has been made to cover the literature published up to December 1968. In addition, there are some more recent references to work published (or in preparation) and available to the author before June 1969, and a few references to unpublished data.

NOTES

Units

Herbage yields and rates of fertilizer application are stated in terms of either lb/acre or kg/ha, the unit employed generally being that given in the original publication.

For conversion : 1·0 lb/acre = 1·12 kg/ha
1·0 kg/ha = 0·89 lb/acre

Terminology

Chemical symbols are used for nitrogen and other nutrient elements.

Quantities and percentage contents of nutrient elements (P, K, etc.) are stated in terms of the actual elements (not their oxides).

The abbreviations DM (dry matter) and OM (organic matter) are used.

The figures quoted for percentage and parts per million contents of herbage constituents refer to contents in the DM, unless otherwise stated.

The term ' response to fertilizer N ' is used to indicate the increase in herbage DM yield obtained from a specified quantity of fertilizer N. Response values are expressed in terms of lb (or kg) DM increase per lb (or kg) N applied, in comparison with control plots not receiving fertilizer N.

PART I

NITROGEN TRANSFORMATIONS IN GRASSLAND ECOSYSTEMS

CHAPTER 1

THE NITROGEN CYCLE AND SOURCES OF NITROGEN

A diagram showing the N cycle in relation to grassland is given in Fig. 1. Different phases of the cycle vary in importance, depending on sward type and management. Factors of major importance are whether the sward has a legume component, whether it is cut or grazed, and whether fertilizer N is applied. Many of the transformations proceed simultaneously and some compete with one another for particular forms of N. As Martin (322) has pointed out, for any one ecosystem the N cycle is best regarded as a flow sheet of possibilities, rather than as a closed system involving the quantitative circulation of N indefinitely.

The atmosphere is the original source of almost all the N in soil/plant/animal systems. Igneous rocks contain only 10-50 ppm. N (143, 436). The transference of N from the atmosphere to the soil is achieved almost entirely by either microbiological or industrial fixation of molecular N_2, though a small amount of chemically combined N is contributed by precipitation.

It is impossible to estimate with any precision the world total for N fixed by micro-organisms, but Donald (143) considers an annual value of about 100 million tons as probable. This compares with the value of 24·1 million metric tonnes (21·9 million tons) N for the industrial production of fertilizer N during the year 1967–8 (160). Industrial production has increased greatly over the past 20 years, being only 2·9 million metric tonnes (2·6 million tons) in 1947–8 (159), and further rapid increases appear likely.

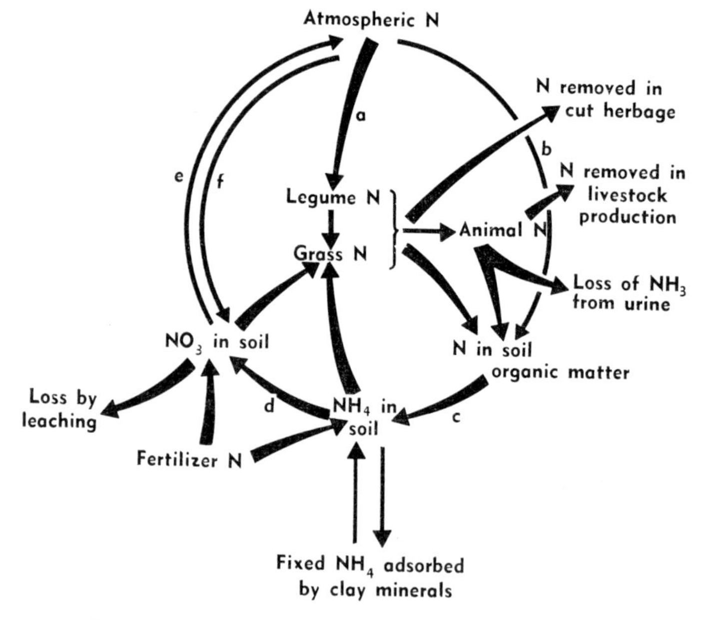

a Nitrogen fixation by symbiotic *Rhizobium* bacteria
b Nitrogen fixation by free-living bacteria
c Ammonification
d Nitrification
e Denitrification
f Additions of combined nitrogen from the atmosphere

Fig. 1. Nitrogen transformations in grassland ecosystems

The quantities of fertilizer N applied to grass swards vary greatly from one part of the world to another and between individual farms in any one area. The wide variation in the average amounts applied in different European countries is illustrated by the following figures compiled by van der Molen (338) for applications to permanent grass:

Netherlands,	132 lb N/acre/year	France,	6·2 lb N/acre/year
W. Germany,	36 ,,	Austria,	4·5 ,,
UK,	19 ,,	Italy,	4·0 ,,

These figures reflect such variables as the reliance on clover N in some areas, the inclusion of mountainous and semi-arid areas, the use made of liquid manure or slurry, and farm size and available capital. Van der Molen pointed out that although the average figure for the UK was 19 lb N/acre, for that grassland which actually received fertilizer N the average rate applied was 76 lb. Further figures for rates of N applied to grassland in England and Wales were given by Yates and Boyd (515) and are shown in Table 1.

TABLE 1

Percentage distribution of fertilizer N application rates on various types of sward in England and Wales in 1962 (Yates and Boyd (515))

Range, lb N/acre/year :	Nil	0–56	56–112	112–168	> 168
Sward type					
Leys (excl. 1-year leys)					
Grazed					
Strip-grazed, %	7	26	42	18	7
Other grazed fields, %	47	35	14	3	1
Mown					
Hay, %	33	51	14	2	0
Silage, %	13	31	38	14	4
Permanent grass					
Grazed, %	71	23	6	0	0
Mown, %	51	37	10	2	0

Preliminary results of a subsequent survey carried out in 1966 showed an increase in the use of fertilizer N on grass, particularly temporary grass, for which the average application was 56 lb/acre per year (53).

CHAPTER 2

UPTAKE AND ASSIMILATION OF NITROGEN BY GRASS

It is thought that non-leguminous plants normally absorb practically all their N as nitrate and ammonium ions. Both are readily absorbed, but the conversion of ammonium to nitrate by nitrifying bacteria in the soil probably ensures that much is taken up in the form of nitrate even when ammonium fertilizers are used. Most plants are thought to take up nitrate rather more readily than ammonium (459), although there is evidence (9) that young seedlings prefer ammonium. A preferential uptake of ammonium was demonstrated by Berlier and Guiraud (38) for young ryegrass plants growing in nutrient solutions labelled with ^{15}N. Culture-solution studies of factors influencing the uptake of ammonium and nitrate both separately and together have been reported by Lycklama (299). Nitrate absorption was found to increase with increasing temperature over the range $5° - 35°$ C and to show a pH optimum at pH 6·2, while ammonium absorption was greatest at a temperature of about 22°C and was not greatly influenced by pH at values over the range $4{\cdot}0 - 7{\cdot}5$. Soluble organic compounds, including amino acids, can also be absorbed by plant roots (311, 330) and, as free amino acids do occur in soils, particularly when microbiological activity is high (377), it is possible that they contribute to plant uptake of N.

Grasses have a very high capacity for absorbing N, the actual rate depending on growth stage and N supply. Wilman (498) reported that Italian ryegrass given 125 lb N/acre absorbed, on average, 6·7 lb N/acre per day in the period 14–21 days after application. Hunt (242), also working with Italian ryegrass, found a similar rate of uptake in the period 16–23 days after the application of 156 lb N. In Wilman's experiment, uptake in the period 14–21 days after the application of 25 lb N/acre was 1·1 lb/acre per day, while in Hunt's experiment

56 lb N produced an uptake of 1·0 lb/acre per day in the period 16–23 days later. Daily rates of uptake for a number of crops listed by Viets (459) include values of up to 3·9 lb/acre per day for maize and lettuce and up to 4·0 lb/acre per day for potatoes.

With uninterrupted grass growth, the rate of N uptake in terms of lb per acre is greatest at the vegetative stage; and the total N yield in the herbage reaches a peak a few days before

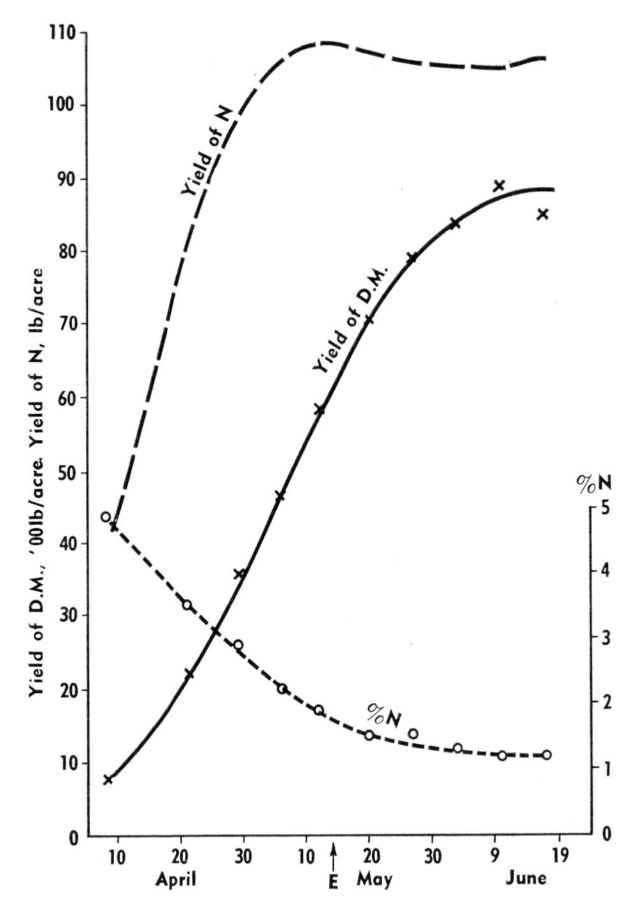

Fig. 2. Dry-matter yield, nitrogen percentage and nitrogen yield in herbage during uninterrupted growth of S24 perennial ryegrass following the application of 54 lb N/acre on 25 February and 70 lb N/acre on 24 March, 1965 (192).

(E=date of mean ear emergence)

the date of mean ear emergence, which is well before peak DM production. The percentage of N in the herbage declines with advancing maturity, as illustrated in Fig. 2. It is probable that some uptake of N continues after the peak yield is attained, but is offset by the death and decay of older leaves.

With a sequence of defoliations, the annual yield of N in herbage will clearly depend on N supply and stage of growth at harvest. Swards receiving an average of 326 lb N/acre per year, applied at the rate of 70 lb for each of 4 or 5 cuts per year for 3 years, gave average annual yields of herbage N ranging from 180 to 238 lb/acre, depending on the grass variety (116).

Once absorbed into the root system, nitrate ions are subject to translocation and metabolism. Some nitrate undergoes reduction in the roots but most is translocated as such to the above-ground parts (459). Reduction of the nitrate to ammonium and its subsequent conversion to protein is normally fairly rapid in leaf tissue, but these processes are related to the rate of photosynthesis. When N uptake is excessive in relation to photosynthesis, there is an accumulation of nitrate when N is absorbed as nitrate, and of amides when N is absorbed as ammonium. These accumulations are increased by deficiencies of nutrients other than N (1).

An outline of the main biochemical pathway by which nitrate is converted into protein in plants is given in Fig. 3. Further details of the enzymic mechanisms involved and of possible alternative pathways are given by McKee (311) and Davies, Giovanelli and Rees (125).

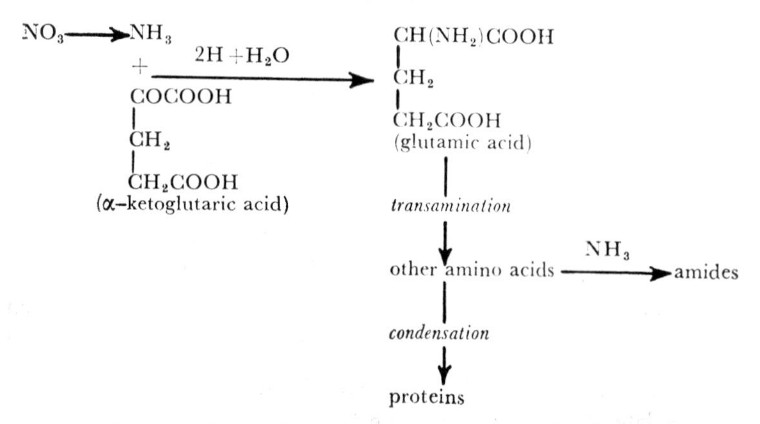

Fig. 3. Conversion of nitrate to protein in plants.

CHAPTER 3

SYMBIOTIC FIXATION BY LEGUMES

Quantities of nitrogen fixed

Light intensity, temperature and rainfall are important in determining the potential for N fixation. According to Martin (325), active fixation requires a temperature of about 48°F. In the favourable conditions of New Zealand, symbiotic fixation has been reported to amount to as much as 600 lb N/acre per year (415, 471). In Britain maximum quantities are probably between 300 and 400 lb/acre.

At Jealott's Hill in S. England, more than 300 lb N/acre has been obtained in the annual production of herbage from irrigated mixed red clover/white clover swards cut 5-6 times per year (248). Calculations from the data given by Cowling (106) show that at Hurley in the wet year of 1958, the N yield of a white clover sward not receiving fertilizer N averaged 283 lb/acre. In W. England, a white clover sward produced an average herbage N yield of 340 lb/acre over 3 years and a grass/clover sward an average of 255 lb (18). In a 3-year experiment at Hurley, various non-irrigated grass/white clover swards gave average yields of 160–200 lb N/acre per year in harvested herbage (116). At Cambridge, lucerne sown at very high density and receiving fertilizer P and K gave yields of N in the harvested herbage of 226, 256 and 306 lb/acre per year respectively for 3 successive years (254). At Aberystwyth, a maximum annual average yield of 268 lb N during 3 years was obtained in the harvested herbage of Du Puits lucerne which did not receive fertilizer N (128). In these investigations additional N would have accumulated in stubble, roots and soil, but this is likely to have been at least partly offset by gains from non-symbiotic fixation and from ammonium- and nitrate-N in the atmosphere. It is probable, therefore, that the herbage yields of N quoted above approximate quite closely to the quantities of N fixed symbiotically by *Rhizobium* bacteria.

In agronomic work, the amounts of N fixed by clover in mixed grass/clover swards have often been assessed in terms of the amounts of fertilizer N required to produce equivalent yields of DM or herbage N from all-grass swards, and values from a number of investigations are presented in Table 2. The actual values obtained will have been influenced by the vigour of clover growth and, where possible, estimates of the contribution of clover to the herbage yield have been included.

The values will also have been influenced by the response to, and percentage recovery of, fertilizer N by the grass swards. Neither set of figures can be regarded as indicating the quantity of N fixed by the clover, and a better estimate for this value is probably given by the yield of herbage N.

Influence of environmental factors

The quantity of N fixed by legumes in the field can be greatly influenced by environmental factors operating on the nodule bacteria, the host plant, or on both (325, 460).

The growth of legumes is generally more influenced than that of associated grasses by soil moisture (339, 438) (lucerne, however, is very drought-resistant), by soil pH (66, 139), and by supplies of P (345, 359, 408), K (51, 189, 297, 298, 318, 343), S (470) and certain trace elements (292). White clover is greatly influenced by weather conditions and Cowling (107) has pointed out that the marked variations in the quantities of N fixed by clover at Hurley are largely a reflection of weather conditions. Thus in 1958, a wet year with vigorous clover growth, cocksfoot alone would have required 228 lb of fertilizer N to produce the same herbage yield as a cocksfoot/white clover mixture receiving no N. In 1956, when the clover contribution was low, the comparable figure was only 65 lb N. Furthermore, Stiles (438) reported that irrigation increased the average quantity of N fixed by white clover in mixture with ryegrass from about 150 lb to about 250 lb N/acre per year. On the other hand, Kleter (271) was unable to detect any correlation between the clover content of permanent grassland and rainfall over a 6-year period. The greater requirements for P and K shown by swards containing legumes often appears to be due to competition from the grass component, rather than to any in-

TABLE 2

Amounts of fertilizer N, in lb/acre/year, required by all-grass swards to replace the effect of clover in grass/clover swards

(Each comparison comprised all-grass swards receiving a range of fertilizer N levels and a grass/clover sward not receiving fertilizer N. White clover was used in the first 9 investigations ; red clover in the last two.)

Grass species	Location	Sward management	Duration of experiment, years	% Clover in herbage yield of grass/clover	lb N required by all-grass sward		Reference
					DM yield basis	N yield basis	
Ryegrass, S23	Hurley, Berks	Cut	3	35–77	171	278	115, 116
,, S24	,,	,,	,,	30–70	148	270	,, ,,
,, Irish	,,	,,	,,	23–71	145	287	,, ,,
Timothy	,,	,,	,,	31–58	120	233	,, ,,
Meadow fescue	,,	,,	,,	32–65	152	261	,, ,,
Cocksfoot	,,	,,	,,	19–58	124	228	,, ,,
Agrostis	,,	,,	,,	60–74	205	298	,, ,,
Cocksfoot	Hurley, Berks	Cut	3	40	160 (65–228)	200	107
Ryegrass, S23	Devon	Cut	2	40–50	180–240	—	420
,, ,,	,,	Grazed	,,	32–37	140–150	—	,,
Mixed	Devon	Grazed	2 and 4	27	90–140	—	64
Various swards	Somerset	Cut and grazed	5	35	133	—	18
Perennial ryegrass	Leeds, Yorks	Cut	3	—	182	265	229
Ryegrass	Ayr	Cut	3	27–44	121	164	391
Cocksfoot	,,	,,	,,	9–22	97	108	,,
Various species grown separately	,,	Cut	3	16–52	123 (41–205)	233 (94–384)	236
Cocksfoot	Beltsville, USA	Cut	2	18	160	200	464, 465
Tall fescue	,, ,,	,,	,,	31	160	240	,, ,,
Cocksfoot } Bromegrass }	Ames, Iowa, USA	Cut	2	40–80	about 240	—	90
3 species grown separately	Helsinki, Finland	Cut	3	—	90	142	385

herently greater requirement by the legumes. Such competition sometimes results in the application of a deficient nutrient being followed by an increase in the growth of grass and a reduction in legume growth, as has been reported for P by Brown and Munsell (66). However, a large response to the combined application of P and S by a grass/clover sward was obtained in New South Wales by Atkinson *et al.* (22) and the N content of the herbage was also increased.

Grass decreases the growth of clover by shading (484) and the vigour of clover growth in mixed swards is therefore influenced by the frequency of defoliation. Cowling (110) found that the N fixed by white clover growing in association with ryegrass was equivalent to 150 lb fertilizer N/acre when the sward was cut 3 times per year, to 180 lb N when cut 6-8 times and to 245 lb N when cut twice weekly throughout the season. Although white clover growing in association with grass is encouraged by frequent defoliation, red clover grown with grass is favoured by a harvesting system involving only 2 or 3 defoliations per year (140).

Karraker *et al.* (264) found that, although legumes fixed less total N per unit area when grown with grass than when grown alone, the amount fixed per unit of legume herbage was much larger. Lucerne, red clover and white clover were all nearly twice as efficient in fixing N when grown with grass as when grown alone.

Reduction of clover growth and of symbiotic fixation by fertilizer nitrogen

The application of fertilizer N to a grass/legume sward usually, though not invariably (62), reduces growth of the legume. This is thought to be due mainly to increased competition from the grass (139, 144, 315, 343, 433, 434, 499). The extent of this reduction depends on such factors as time of N application, frequency of sward defoliation, supplies of moisture and nutrients, and the form in which the N is applied.

Results obtained by Young (516) indicated that N had least effect on clover if applied in early spring and autumn when clover growth was not very active. However, Cowling (108) considered that the effect of N was reduced where clover stands were well developed and that applications of N to such swards in June might cause no more suppression than dressings in March/April.

Frequent defoliation of swards, especially if to a height of about 1 in., tends to offset the effect of applied N on white clover by reducing the shading caused by the grass (45, 230, 401, 432). In experiments carried out by Robinson and Sprague (40) in the N.E. States of the USA, clipping to 0·5 in. when the herbage was 4 in. high maintained a clover content of about 28% even when 300 lb N/acre per year was applied to a non-irrigated sward. However, Holliday and Wilman (230) found that even with 10 cuts per year, clover contributed only 3·8% of the herbage yield where N was applied at 139 lb/acre per year; with two cuts, the contribution was 0·1%. Without applied N, the herbage contained 41% clover with 10 cuts and 5·8% with 2 cuts.

The effect of irrigation on the suppression of clover by fertilizer N varies very considerably with the climate. In S.W. Scotland, Reid and Castle (390) found that irrigation had little effect on the clover content of mixed swards receiving various rates of N up to 312 lb/acre per year over a 3-year period. However, experiments in S.E. England (294, 378, 438) and in Ohio (382) have indicated that irrigation offset to some extent the suppression of clover by applied N. Values obtained by Low and Armitage (294) are given in Table 3:

TABLE 3

Percentage of clover in the herbage yields of perennial ryegrass/white clover swards as influenced by N supply and irrigation (Low and Armitage (294))

Total N applied before the 4th cut	Nil	30 lb/acre	60 lb/acre
Date of 4th cut	9 Sept.	26 Aug.	18 Aug.
	% Clover		
No irrigation	52	8	4
Irrigated at 2·00 in. deficit	60	14	3
,, ,, 1·34 in. ,,	65	21	8
,, ,, 0·67 in. ,,	71	28	10

In other areas, even greater effects have been obtained. Thus Robinson and Sprague (401) found in the N.E. States of the USA that with irrigation, N applied at about 400 lb/acre per year depressed clover content only from 50 to 42%, whereas without irrigation corresponding figures were 45 and 24%. In these trials the herbage was cut to 1 in. when 4–5 in. high. In an experiment in South Australia, white clover also contributed 42% of the herbage of an irrigated mixed sward receiving 295 lb N/acre per year (263).

Applied K has sometimes reduced the clover-suppressing effect of N (51, 297). Such effects would be expected only on K-deficient soils. However, in trials in Scotland, Reith *et al.* (396) noted that whereas fertilizer K usually increased the clover contents of swards in the absence of fertilizer N, its effect was negligible when N was applied. Subsequent trials by Lowe (298) on a K-deficient soil also indicated that fertilizer K could not prevent the depression of clover when 190 lb N/acre per year was applied.

Not all forms of N have the same effect on the clover content of mixed swards. Moloney and Murphy (339) noted that clover was depressed more by ammonium sulphate than by Nitro-chalk (a mixture of ammonium nitrate and calcium carbonate), an effect probably due to increased soil acidity brought about by the ammonium sulphate. Urine and liquid manure often contain high N concentrations and are reported to have reduced the clover content of mixed swards in some investigations (223, 407, 456, 480), and to have increased it in others (48, 92, 153, 179). In experiments reported by Castle and Drysdale (92), mean clover contents of swards which received (a) liquid manure, (b) equivalent amounts of fertilizer N, and (c) no fertilizer, were 32, 18 and 15%, respectively. In subsequent work reported by Drysdale (153), liquid manure (from a cow-shed), solid fertilizer and aqueous fertilizer solutions, all containing the same weights of N and K, produced similar herbage yields. However, liquid manure resulted in consistently higher clover contents in the sward at N levels up to 261 lb N/acre per year. Drysdale also found that clover contents were high where both liquid manure and fertilizer N were applied together. Thus, at a rate of 200 lb N/acre, swards receiving liquid manure, fertilizer, or a mixture providing half the N in each form, contained 20%, 8% and 18% clover, respectively. Total yield for each of the 3 treatments was about 10,000 lb DM/acre. Thomas (451) also stated that where stocking rate is high and also where animal excreta are returned as slurry, clover can flourish even though very large amounts of fertilizer N are applied. The reasons for the different effect of liquid manure and fertilizer N are not clear and require further investigation. In Drysdale's work the effect did not appear to be due to the supply of either K or water.

Although fertilizer N applied to grass/legume swards almost invariably depresses clover growth, the extent to which N fixation by the remaining clover is reduced is uncertain. When growing alone, legumes absorb mineral N with a corresponding reduction in N fixation. This has been demonstrated for white clover by Allos and Bartholomew (15) and Walker *et al.* (471), and for lucerne and Ladino clover by McAuliffe *et al.* (307). Furthermore, Cowling (106) found that 100 lb N applied to a pure white clover sward almost halved the weight of root nodules. However, when growing in association with grass, legumes appear to take up only a small proportion of any N applied as fertilizer. In a pot experiment, Walker *et al.* (471) found that when grass and white clover were grown together, the clover absorbed only 5–6% of the applied N over a wide range of application rates, and calculations based on herbage N yields from field experiments have indicated that fertilizer N applied to a mixed sward does not impair the ability of the remaining clover to fix N and transfer it to the associated grass (64, 91, 107, 236). In fact, Young (516) reported that 40 lb N applied to a mixed sward in early spring appeared to have a slight beneficial effect on clover growth. McAuliffe *et al.* (307) found that seedlings of white clover and lucerne were unaffected by inorganic N, but there was a marked depression of N fixation in 10-week-old plants and in established plants.

It is clear that, in practice, the quantities of N fixed symbiotically in grassland swards vary enormously. The potential fixation for any site is largely governed by climatic factors, especially temperature and water supply. The proportion of the potential actually achieved depends mainly on sward management. Inadequate supplies of nutrients other than N generally restrict the growth of clover more than that of grass; and the application of fertilizer N almost invariably reduces the clover content of mixed swards and hence reduces N fixation per unit area.

CHAPTER 4

TRANSFER OF SYMBIOTICALLY FIXED NITROGEN TO GRASS

The presence of clover in swards not receiving fertilizer N often increases not only the total herbage yield, but also the yield of the grass (221, 412, 420). This increase in grass growth is due to the transfer of N from the clover and occurs despite competition for light, water and nutrients.

When swards are grazed, much of the N in the herbage consumed is returned in excreta. It is therefore theoretically possible for much of the N fixed by clover to be returned to the soil and be absorbed by the grass component of the sward. Transfer of N also occurs in non-grazed swards, indicating that it is released from dead and/or living legume tissue (221, 473). The decomposition of root and nodule tissue is probably the main route for this, but there is evidence that N may be transferred via the excretion of organic compounds from living roots. In addition, some N undoubtedly reaches the soil in dead leaves and stems, and possibly also by leaching from leaves (455).

Several workers consider that in grazed swards, the cycling of N through livestock is probably the most important means of transfer from clover to grass (311, 403, 411, 473). However, there is evidence (see Chapter 6) that the N returned in animal excreta is very poorly utilized by grass swards owing to its uneven distribution and to losses caused by volatilization and on some soils by leaching. Shaw *et al.* (420) have attempted to assess the relative quantities of N transferred from clover to grass in both cut and grazed swards. Although grazed swards contained about 50% less clover than cut swards at all levels of fertilizer N, the transfer of N was greater. The data in Table 4 show results obtained in the absence of fertilizer N. Allowing for the clover content of the grazed swards by comparing the N transference per lb clover N, it appears that the

TABLE 4

Quantities of N transferred from clover to grass in grass/clover swards (Shaw *et al.* (420))

Year	N transferred, lb/acre		lb N transferred/lb clover N	
	Cut	Grazed	Cut	Grazed
1963	38·6	59·8	0·20	0·45
1964	70·6	79·2	0·37	0·56

return of excreta by grazing approximately doubled the quantity of N transferred. However, as the N yield in the clover component of the grazed swards was 133 and 141 lb/acre in 1963 and 1964, respectively, and since excretion of N could be expected to amount to at least 75% of that ingested, it is clear that much of the returned N was not recovered. This is consistent with other work (see Chapter 6). Earlier work reported by Brockman and Wolton (64) indicated that while the N transfer in mown swards remained fairly constant throughout the season, on grazed swards it increased steadily from May to September, indicating that the return of excreta had a cumulative effect.

The transfer of N in the absence of excreta may amount to more than 50% of that fixed (467), but is often much less. Several workers have reported that little transfer takes place in the establishment year of a grass/white clover ley (229, 473), presumably because the clover is itself using all the N fixed and there is little root decomposition. Cowling *et al.* (114) showed that N transfer to grass in a cut grass/clover sward not receiving fertilizer N amounted to 28–65% of the total N present in the mixed herbage and that transfer was considerably higher in the 3rd year than in the 1st two years. The values for swards receiving fertilizer N are less reliable. The experiment reported by Bland (49), in which the effect of separating the roots of perennial ryegrass and white clover was examined, also indicated that transfer was insignificant until the 3rd year. The data of Herriott and Wells (223) for plots not receiving either fertilizer or animal excreta indicate that in 4 successive years, N transfer calculated on a herbage basis amounted to 0, 54, 68 and 76% of that fixed, while the proportion of clover in the herbage fell from 49 to 12% over the same period. The decomposition of clover roots and herbage presumably contributed largely to the transfer.

Reid and Castle (391) estimated that white clover transferred 36% of its fixed N to ryegrass and 51% to cocksfoot over a 3-year period in cut swards. The actual quantities transferred in the two swards were 39 and 36 lb N/acre, respectively. In a 3-year experiment carried out by Cowling (107), the average N transfer from white clover to cocksfoot was 60 lb/acre per year, and in one year there was an apparent transfer of 106 lb. The yield of clover herbage was high in this experiment, being almost twice that in Reid and Castle's experiment. The difference in clover yield and N transfer between these experiments may have been partly attributable to differences in clover seed rate. Herriott and Wells (221) reported an increase in N transfer as a result of increasing the seed rate of clover from 1 to 3 lb/acre. Although it had little effect on the yield of clover herbage, the higher seed rate was thought to have increased the root mass and hence the amount of N released on decomposition.

Experiments on the mode of underground transfer of N and factors which affect its magnitude have indicated that the release of N from decomposing root material is important. In a pot experiment, Simpson (424) found that white clover competed with grass for N until the autumn/winter period, when some transfer, probably due to root decomposition, occurred and that subterranean clover, an annual species, did not release any N until senescence and then did so rapidly. Butler *et al.* (83) and Wilson (501) found that with white clover, repeated defoliation caused a rapid turnover of root and nodule tissue, involving death of the older material and its replacement by new roots which soon became heavily nodulated. With red clover (83), loss of roots and nodules was less marked, and repeated defoliation resulted in progressively poorer growth. Dilz and Mulder (139) also reported that although legumes made a small contribution to the N supply of associated grasses during active growth, there was a much greater release of N after the plants were killed by complete removal of the herbage. With white clover, but not with red clover or lucerne, normal cutting treatments markedly increased N release, presumably owing to decomposition of part of the root system. Shading also caused a loss of root and nodule tissue in white clover, and to a lesser extent in red clover (83, 441). These results indicate that conditions in rotationally grazed or regularly cut swards favour the transfer of N from clover to grass by turnover of root and nodule tissue.

There has been some controversy regarding the importance of the excretion of N compounds from living legume roots. Virtanen and co-workers in Finland demonstrated (see Wilson, 502) that excretion did occur from the roots of various legumes and that this N could be utilized by associated non-legumes. Workers elsewhere, however, were unable to detect any appreciable excretion from these species, which included lucerne, and it therefore appeared that such excretion occurred only when long days were combined with rather low temperatures, so that organic N compounds of low molecular weight tended to accumulate in the nodules and roots (see 82, 502). More recently, however, in a pot experiment at Canberra in Australia, Simpson (424) found that lucerne released N gradually over the whole 18-month period of his experiment and pointed out that this was more consistent with an excretion mechanism than with the decomposition of root material. In this experiment, frequent defoliation of the lucerne reduced the amount of N transferred. Figures for N transfer from lucerne and white clover obtained by Simpson were lower than others reported from the Netherlands, and he suggested that this may have been due to the higher temperatures and light intensities combined with shorter daylengths in Australia. Henzell (219) also obtained very little N transfer from lucerne and white clover to grass under the subtropical environment of Brisbane.

Although the quantities involved are hard to determine under field conditions in mixed swards, there appears to be considerable transfer of N from the decay of unharvested leaf and stem material. Cowling (106) found that after cutting an all-clover sward with hand shears, the N remaining in the unharvested stolons and petioles amounted to 133 lb/acre, compared with 47 lb/acre in the underground organs. Although there is little published information on the rate of turnover of clover stolons, Bakhuis and Kleter (27) demonstrated that N transfer did occur through the decomposition of leaf and stem material and/or by the leaching of N from leaves. In their experiment, the N uptake of grass grown normally in alternate rows with white clover was compared with that where the roots of the two species were separated by vertical plates 60 cm deep. Grass and clover each grown alone were used as controls. In the establishment year, there was little above-ground transfer of N, but in the following year above-ground and below-ground transfer were approximately equal in magnitude. The importance of the decomposition of above-ground plant material is confirmed by the large seasonal fluctuations in the weight per unit area of white clover stolons reported by Vez (458). Working in Zurich, he found reductions of 30–56% during the winter and smaller reductions following defoliation.

The work summarized in this chapter suggests that the percentage of the N fixed by the legume component of a mixed sward that is transferred to the grass component may vary from nil to 75%. There is normally very little transfer in the year of establishment and maximum rates are unlikely to be attained before the 3rd year. Transfer from plants more than one year old depends on the frequency and height of defoliation, the presence or absence of the grazing animal, and climatic factors which govern the balance between N fixation and photosynthesis.

CHAPTER 5

CONSUMPTION OF NITROGEN BY RUMINANTS AND ITS REMOVAL IN LIVESTOCK PRODUCTS

Herbage from cultivated grassland is used almost entirely as a feed for ruminant animals. Of the N consumed, part is converted by the animal into body tissue or into milk and the remainder is excreted.

Ruminants have the characteristic feature of being able to digest appreciable amounts of cellulose and hemicellulose. The rumen micro-organisms involved in this process also participate to a large extent in the conversion of the N compounds in the herbage into animal protein. Several reviews (43, 312, 371) have dealt with the transformations of N within the rumen.

The N requirements of ruminants have been reviewed by a Technical Committee of the Agricultural Research Council, UK (4). This Committee considered that all diets for ruminants should contain at least 1·4% N (9% crude protein), even if this level exceeded the calculated minimum requirement based on the production of protein in milk, liveweight gain, etc. The level of 1·4% N appeared to be necessary for adequate activity of the rumen micro-organisms. For highly productive animals, a dietary content of 2·2–2·4% N (14–15% crude protein) has been considered generally adequate (388).

Of the N present in grass and legume herbage, about 75–90% is usually present as protein and the remainder mainly as peptides, amino acids and amides, although nitrate can be an important constituent in some situations (see pp. 135–7).

In exceptional circumstances, animals can suffer from toxic effects induced by N compounds in the diet. With herbage containing high levels of nitrate there is a risk of toxic levels of nitrite occurring in the rumen (see p. 152). Such levels are likely only when microbial activity is poor, since in normal circumstances the further reduction of nitrite to ammonia is rapid (371).

The removal of livestock products results of course in the withdrawal of N from the soil/plant system, but the quantities involved are small in relation to those recommended for fertilizer application. Under fairly intensive levels of grassland management, it is possible to produce, per acre per year, 1000 gallons of milk, 800 lb liveweight gain in beef cattle, or 1000 lb liveweight gain (including fleece weight) in sheep plus lambs. Assuming these levels of production, and N contents of 0·64% for milk and 2·4% for liveweight gain in cattle and sheep (4), removals of N per acre per year would amount to 64 lb for dairy cattle, 19 lb for beef cattle and 24 lb for sheep and lambs.

CHAPTER 6

RETURN OF NITROGEN IN ANIMAL EXCRETA

The proportion of N consumed which is subsequently excreted and its partition between urine and faeces depend on the type of livestock and the N content of the diet. Walker *et al.* (473) stated that beef cattle excrete about 95% of the N ingested and dairy cattle about 75%. According to Barrow (31) and Barrow and Lambourne (33), both cattle and sheep excrete in the faeces a fairly constant 0·8 g N per 100 g DM consumed. The remainder of the N is excreted in the urine, the proportion depending on the N content of the diet. With herbage containing more than 4% N, Barrow and Lambourne (33) found that about 80% of the N excreted by sheep was present in the urine. With herbage containing about 0·8% N, the proportion of the excreted N present in the urine was only 43%. Other published data are consistent with this general picture. Thus Sears *et al.* (416) and Sears (410) reported that with sheep grazing grass/clover swards in New Zealand, the urine contained 70–75% of the excreted N. With cattle grazing grass swards in the USA, Lotero *et al.* (293) considered that the N excreted was equally distributed between urine and faeces; and with cattle grazing native pasture in South Africa, Gillard (178) reported that the faeces contained more than 80% of the excreted N.

As an example of the quantities of N involved in practical situations, dairy cattle consuming 10,000 lb of herbage DM containing 2% N will excrete the equivalent of about 150 lb N/acre per year.

Of the N in urine, about 90% occurs as urea or amino-N, and although readily available for plant and microbial uptake, this is also subject to losses. The N in faeces is in insoluble forms and depends largely on the activities of the soil fauna for incorporation into the soil.

Return of nitrogen in urine by grazing animals

The N content of urine is very variable, although reports of average values agree quite closely. For cattle the usual range seems to be 2·5–13 g N/l., with an average of about 8 g/l. (39, 47, 142, 154, 179, 380). For sheep, reported values vary from 5·7 to 14·7 g/l., with an average of about 9 g/l. (121, 142, 223, 417).

The frequency of urination in cattle is generally about 8-12 times per day (206, 316, 380), and of sheep about 24 times per day (142). Average volumes per urination reported for cattle include 1·6 l. (probably for Channel Island cattle, although this is not stated) (142), 2·2 l. (126) and (for Ayrshire cows) 2·5–3·5 l. (47). For sheep, Doak (142) reported an average volume of 150 ml. There is however considerable variability in urine production. Vercoe (457) reported from Australia that the daily volume for sheep varied from 2·8 l. in June to 0·5 l. in December.

The ground area covered by a given volume of urine is considerably influenced by soil texture and moisture status. With cattle, Davies *et al.* (126) reported an area of only 2 ft^2 as being covered by 2·2 l. urine; Petersen *et al.* (380) found that each urination covered an average of 3 ft^2, and Doak (142) stated that 1·6 l. urine added to bare, firm ground artificially at approximately the normal rate, wetted an area of 4 ft^2. With sheep, Doak found that 150 ml urine covered 45 in.^{2} of bare soil surface, which is certainly greater than would have been covered on a sward. Assuming for cattle a grazing intensity of 200 cow-days/acre per year, an average area wetted of 3 ft^2 and 10 urinations per day, then about 14% of the area will receive urine per year, assuming that no overlapping of urine patches occurs. With intensive sheep grazing, Herriott and Wells (223) calculated that on the basis of a urination volume of 150 ml and a ground coverage of 45 in.^{2}, 27% of the area would receive urine in any one year.

The average rate of N applied to an area covered by a cattle urination of 2 l. containing 8 g N/l. and distributed over 3 ft^2, would be equivalent to 514 lb/acre; and for a sheep urination of 150 ml containing 9 g N/l. and covering 45 in.^{2}, 377 lb/acre. However, the centre of the urine patch will receive more and the periphery less N per unit area than the overall average amount.

Owing to the lateral spread of roots and diffusion of N in the soil, the influence of a urination will extend over a greater area than that actually wetted. Doak measured the areas of stimulated growth on a sward grazed by sheep and obtained a mean value of 100 in.^{2} for a large number of urine patches, or slightly more than double the area of bare ground actually wetted. Assuming that the patches were circular, growth was affected to a distance of about 2 in. from the edge of the patch. With cattle, Lotero *et al.* (293) reported that the area influenced was generally 9·5–13 ft^2. and Blagden (47) found that the area on which growth was stimulated by 3·5 l. of urine applied artificially varied from 6 to 23 ft^2.

The fate of the N in urine is very much influenced by weather and soil conditions, but some loss as gaseous ammonia seems to be inevitable. Doak (142) reported that urine applied at 2 l. per 10 ft^2 raised the pH of the soil surface from 5·3 to 7·8 within 4 hours. This increase in pH would increase the volatilization of ammonia. In another experiment in which the maximum air temperature was 21°C and no rain fell on the urine patch, Doak found that 12% of the N was liberated as gaseous ammonia within 3 days. He considered that this 12% loss might be considerably exceeded in some situations, although rain could reduce it. In investigations in which sheep urine was applied to lysimeters cropped with ryegrass, Watson and Lapins (477) found that about 50% of the N was lost over a two-year period through causes other than leaching, presumably mainly by volatilization of ammonia. A small-scale investigation at Hurley indicated that there was a loss equivalent to 6 lb N/acre per hour from urine patches, and that half the N could be recovered from the atmosphere as ammonia within two or three days during warm, sunny weather (183).

In addition to losses as gaseous ammonia, other gaseous and leaching losses are likely to occur to a greater extent from urine patches than from the bulk of the soil. Laboratory studies by Doak (142) indicated that, with relatively acid soils, nitrite could accumulate in a urine patch and, as described on p. 50, this may result in a loss of gaseous N. High levels of N also increase denitrification (see Chapter 12). Although leaching of N would not be expected during the growing season, it might well occur from urine patches during winter. Evidence of such leaching has been reported from New Zealand (see p. 47).

Return of faecal nitrogen by grazing animals

The quantity and composition of faeces vary with diet, but the average N content is about 0·4% on a fresh-weight basis, or about 2·0–2·8% on a DM basis (39, 178, 380). The N is of value to the sward only after incorporation into the soil, and considerable losses of gaseous ammonia can occur from faeces which remain on the surface. Gillard (178) reported a loss of about 80% of the N from faeces dried by the sun in South Africa. The rate at which faeces are incorporated into the soil is very much influenced by weather conditions and the activity of the soil fauna (31, 178, 413).

Petersen *et al.* (380) estimated that, on average, the faeces of cattle supplied the equivalent of 760 lb N/acre to the area actually covered, and that at a stocking rate of 200 cow-days /acre, 5·5% of the area would be covered each year.

Estimates of the extent to which the N in faeces affects grass growth in the surrounding area vary. Petersen *et al.* (380) working in North Carolina, considered that the area affected was only slightly greater than the area covered, whereas in the wetter climate of Scotland, MacLusky (316) reported that faeces affected the growth of herbage over an area about 6 times that actually covered.

Return of nitrogen in liquid manure and slurry

The excreta produced by housed stock may be (a) removed as waste, (b) converted into farmyard manure, or (c) collected as liquid manure (urine diluted with water) or slurry (urine with faeces and some water) and applied to the land. Farmyard manure is usually applied to arable land, and liquid manure and slurry to grassland.

The composition of liquid manure and slurry depends mainly on the quantity of water added and the length and conditions of storage. Excreta returned in these forms can be applied much more evenly than those returned during grazing, and at lower rates of application per unit area. Losses of N should, consequently, be reduced and there is some evidence that this is so. However, substantial leaching and gaseous losses may occur from liquid manure and slurry applied during the winter, when uptake by grass is slight.

Utilization by the sward of nitrogen returned in animal excreta

Evidence on the effectiveness of returned excreta in increasing herbage yields is conflicting, though lack of consistency between the results of the various experiments reported is, perhaps, to be expected in view of the many factors which can influence both losses of N and sward response. Among factors affecting losses, uniformity of application and time of year are important. When the response to N in excreta is assessed by comparing cut and grazed swards, the effects of trampling and differences in the pattern of defoliation may influence the comparison. Whether clover is present is also important. With grass/clover swards, various workers have reported that the return of excreta during grazing has failed to increase yields, apparently through suppression of N fixation by the clover (163, 223, 418, 475, 480, 511). In some trials with grass/clover swards, the return of excreta has given a yield response which was at least partly attributable to K (223, 480). In contrast, a trial with a grass/clover sward at Palmerston North, New Zealand, showed an annual yield response of about 30% to return of excreta on a high-K soil, a result presumably attributable to N (416, 417).

With swards containing little or no clover, the return of excreta during grazing has more often given an appreciable, though variable, yield response (18, 411, 412, 475, 480, 511). However, Bryant and Blaser (72) obtained higher yields of cocksfoot with cutting than with grazing; with the same frequency of defoliation, cut swards yielded about 35% more than did grazed swards receiving urine but which had faeces removed after each grazing.

With liquid manure and slurry applied under suitable weather and soil conditions, the effectiveness of the applied N can be as high as with inorganic fertilizers. In experiments on grass/clover swards in Scotland (92, 152), DM yields where up to 400 lb N/acre per year was applied either as liquid manure or as fertilizer were almost identical. Other data on responses to the N in liquid manure and slurry are given on pp. 99–100.

The uneven distribution of animal excreta during grazing and the large losses of N so returned, commonly more than 50%, explain why, in grazing situations, the N cycle is not a closed system with N circulating indefinitely between soil, plant and animal.

CHAPTER 7

NON-SYMBIOTIC NITROGEN FIXATION

That various free-living soil micro-organisms fix atmospheric N is well established. However, the amounts of N involved appear usually to be small and virtually impossible to assess accurately under field conditions (324, 341).

To attribute net changes in soil N content under grass swards entirely to non-symbiotic fixation may involve errors resulting from N additions as dust, ammonia and nitric acid from the atmosphere, or losses through denitrification or leaching (341).

A large gain of just over 100 lb N/acre per year in soil N content was reported from plots of bluegrass without legumes during an 8-year period following 35 years under arable cultivation (486). In 6 of the 8 years, the herbage was cut and removed, but in the other two years it was cut and left on the plots ; no assessment of additions from the atmosphere was made in this experiment. Karraker *et al.* (264), using lysimeters, found that although there was virtually no change in soil N content under a bluegrass sward, there was a net gain to the system of about 33 lb N/acre per year over an 11-year period. This gain was additional to the N gained in precipitation and was considered to represent non-symbiotic fixation. In Western Australia, Parker (373) reported net gains of at least 55 lb N/acre per year in a grass sward without legumes. Of this quantity, about 20 lb occurred in the soil (a heavy loam) and the remainder in the herbage.

Since non-symbiotic fixation requires the presence of readily available sources of energy and is encouraged by a low oxygen partial pressure, Jensen (258) considered that the process is likely to occur to a greater extent in grassland than in arable soils. This view is supported by van Schreven (409), and by Parker (373) who has pointed out that conditions in many grassland soils are likely to favour N fixation by the anaerobic *Clostridium*.

On the basis of laboratory experiments with N-fixing micro-organisms, particularly *Azotobacter*, the fixation of 20–100 lb N was estimated to require a minimum consumption of 1100–5500 lb of " first-class " organic matter (OM), i.e. of the same energy value as glucose (258). A higher ratio of about 100:1 for the ratio between carbohydrate used and N fixed is regarded as normal by van Schreven (409).

Although 4000–9000 lb OM/acre may well be decomposed each year under a grass sward (see pp. 31–2), the quantity available to support non-symbiotic fixation would be much smaller. The conversion of the N in plant and animal residues to microbial and soil organic N would itself involve the consumption of much of the carbonaceous material present. Only materials with C:N ratios wider than about 30:1 are likely to provide appreciable energy for non-symbiotic fixation. This probably explains, at least in part, the finding of Delwiche and Wijler (131) that the incorporation of dried grass material, including roots, at rates of 10, 20 and 40 mg per g soil (respectively equivalent to about 8, 16 and 32 tons/acre to a depth of 6 in.) resulted in very small increases equivalent to only 0·59, 0·81 and 2·4 lb N/acre during incubation for 40 days at 21°C. The incorporation of 10 mg glucose, however, resulted in fixation equivalent to 40 lb N/acre. On the other hand, it is possible that localized zones rich in carbohydrate may occur in the field and result in greater fixation than occurs in artificial mixtures of soil and plant material.

There is considerable evidence that conditions of limited aeration increase the efficiency of N fixation by free-living micro-organisms (258). This is thought to be due to an increased production of organic metabolites from cellulose and their more efficient utilization by N-fixing organisms, including *Azotobacter*. Barrow and Jenkinson (32) showed that when cellulose materials were added to soil, a certain degree of anaerobiosis was necessary to bring about active fixation, but that no fixation occurred unless oxygen was present in the atmosphere above the soil, indicating that the process was not wholly anaerobic. The practical importance of fixation by *Clostridium*, which is more widespread than *Azotobacter* and which grows only in anaerobic conditions, is still uncertain (258).

The application of fertilizer N is likely to restrict non-symbiotic fixation, since the process has been shown in culture experiments to be reduced markedly by the presence of ammonium and nitrate (461). Reductions in *Azotobacter* numbers in soils receiving fertilizer N have also been reported in a number of experiments (see 258). However, assessment of the extent of such reduction in the field is virtually impossible.

From the work summarized above, non-symbiotic fixation appears unlikely to be significant in swards receiving moderate to heavy applications of fertilizer N. However, in swards not receiving fertilizer N, it may contribute amounts of the order of 20–50 lb N/acre per year.

CHAPTER 8

ACCUMULATION OF NITROGEN IN SOIL ORGANIC MATTER

When arable soils are sown to grass, contents of OM and N increase and continue to do so for many years if the sward remains unploughed. The increase is asymptotic, but equilibrium is eventually reached when additions of OM and N are balanced by mineralization and losses. The time taken to reach equilibrium varies considerably with soil type, climate and sward management. The situation at equilibrium can be expressed as $S = A/r$, where S = the total amount of OM (or N) present, A = the annual addition and r = the fraction of the total mineralized each year. Before equilibrium is reached, S will be less than A/r and will therefore increase, but at a diminishing rate, as the equilibrium position is approached.

Data obtained in the UK by Richardson (399) indicated that the attainment of equilibrium under Rothamsted conditions required more than 100 years, and that there was an average increase of about 50 lb N/acre per year during the first 40 years after prolonged arable cultivation ceased. These conclusions are supported by the values of 0·3–0·7% N (equivalent to 6000–14,000 lb N/acre) commonly present in the top 6 in. of permanent pasture soils, and by the rates of accumulation of 57–130 lb N/acre per year over a 3-year period reported by Clement and Williams (100) at Hurley. In their investigation, increases were greatest under grazed ryegrass/white clover swards and least under cocksfoot with all herbage cut and removed. A comparable arable soil showed a constant N content of about 0·12% in the top 6 in.

Work in New Zealand on grazed grass/clover swards indicated an average increase in soil N content of 100 lb/acre per year (468). With grass/clover swards on soil initially very low in OM, Sears *et al.* (415) reported values of more than 150 lb/

acre where the herbage was cut and removed and more than 250 lb/acre per year when the herbage was returned to the soil. From South Australia, Russell (402) reported an average increase of 46 lb N/acre per year in plots of native pasture improved by fertilizer P and grazed by sheep over 30–40 years. The increases reported for previously arable soils in Victoria, Australia, by Mullaly *et al.* (346) are equivalent to 28–94 lb N/acre per year for legume pastures and 19–58 lb for non-legume pastures.

Sources of soil organic nitrogen

It is difficult to assess the quantities of plant residues produced by a grass sward during the course of a year. Both roots and herbage are continuously replenished and contribute to the soil OM, and animal excreta are an additional source of OM on grazed swards. Fertilizer N may be converted into organic form as a result of its uptake and assimilation either by plants or directly by micro-organisms.

There are few reports on the quantities of herbage litter reaching the soil surface, which will, of course, vary with the frequency and method of defoliation. Hunt (244) estimated that, in New Zealand during the winter, 8 lb DM/acre per day was lost from an undefoliated Italian ryegrass sward; this amount, equivalent to 2900 lb/year, would almost certainly have been exceeded during the summer months. The weight of stubble left after mowing a grass sward to a height of 2 in. may be as much as 5500 lb/acre (99) and, as the average life of a tiller is considered to be less than a year, an estimate of 5000 lb/acre per year for dead herbage and stubble from swards yielding about 10,000 lb harvested DM/acre per year seems fairly conservative. The same proportion, 3500 lb of litter from a pasture producing 7000 lb of herbage consumed, was considered an average value for Australia (28). Evidence of a similar turnover of herbage material in a white clover sward has been reported by Vez (458) working in Zurich, who found reductions in stolon weight of 30–56% in winter, and smaller reductions after defoliation.

The N content of dead grass herbage appears to vary from about 0·7 to about 3·4%, depending on N supply during the previous growth period, while the N content of stubble varies from approximately 1·0 to 2·0% for single applications of fertilizer N at rates up to 350 kg/ha (113). Assuming an average N content of 1·5%, dead herbage and stubble amounting to 5000 lb/acre per year would contribute 75 lb N to the soil OM.

The minimum annual DM contribution of roots in a productive sward appears to be about 3000 lb/acre (25), with 4000–4500 lb/acre a probable average value (180). Barley (28), however, considered that only about 1400 lb root material per year would decompose under a pasture yielding about 7000 lb DM. The N content of grass roots is normally 1·0–1·8% of the OM (see Table 5). Thus, assuming an annual root turnover of 4000 lb and a N content of 1·5%, the soil OM would receive 60 lb N/acre from this source. If one-quarter of the root turnover were composed of clover roots containing 3·5% N, the total contribution of N from roots would amount to 80 lb/acre.

The estimates given above for dead herbage and root material amount to 135–155 lb N/acre per year for productive swards in lowland areas, and should, of course, be reduced where herbage yields are less than 10,000 lb harvested DM/acre per year. Grass swards supply greater quantities of residues to the soil than most arable crops, normal figures for cereal roots being 1200–2800 lb/acre, and for sugar-beet root residues only about 500 lb DM/acre (104).

When swards are grazed, animal excreta provide an additional source of OM and N (see Chapter 5), though in practice this has not always increased the rate of accumulation of soil N. Sears and Evans (414) found that after 4 years, the soil N content to a depth of 2 in. was about 25% greater on both grass and grass/clover swards where excreta had been returned. Wolton (509) also found that two years after sowing a grass/clover ley, the return of dung and urine together (though not individually) caused a small increase in total soil N. Clement and Williams (100) reported that the soil N increase in the top 15 cm was 0·019% N after 3 years (equivalent to 129 lb N/acre per year) under grazed grass/clover swards and 0·014% (89 lb) under cut swards. However, Metson and Hurst (327) found no appreciable difference in the total N content of the top 3 in. of

TABLE 5

Carbon and nitrogen contents and C : N ratios of the roots of grasses and legumes

Material	%C	%N	C : N	Reference
GRASSES				
Ryegrass roots (living roots from plants grown in gravel with nutrient solution)	47*	1·03	45·6	442
Bromegrass unfertilized	—	0·84	—	381
Bromegrass fertilized with 200 ppm. N in soil	—	1·44	—	381
Bromegrass (crown and root material)	36·2	1·16	31.2	149
Ryegrass roots (living roots from sward ; average for monthly samples over 18 months)	47·6†	1·50	31·7	170
Ryegrass roots (living and dead from sward ; average for monthly samples March –June)	50·6	1·75	28·9	484
Cocksfoot roots (ditto)	49·2	1·54	31·9	,,
Timothy roots (ditto)	48·9	1·47	33·2	,,
Ryegrass roots (living and dead from soil) :				
no fertilizer N in season	49·4	1·07	45·8	,,
200 kg N/ha in previous 17 weeks	48·4	1·22	39·6	,,
400 kg N/ha in previous 17 weeks	48·5	1·64	29·7	,,
650 kg N/ha in previous 17 weeks	48·1	1·92	25·0	,,
Roots of 3 South African species :				
unfertilized	—	0·37–0·74	—	478
fertilized with 126 lb N	—	0·45–1·04	—	,,
Roots of various grasses	—	0·46–0·76	—	422
LEGUMES				
Lucerne (crown and root material)	42·2	1·98	21·3	149
Clover roots:				
few-months-old sward	—	2·99	—	275
2-year-old sward	—	2·09	—	,,
White clover roots (living roots from mixed sward ; average for monthly samples over 18 months)	47·6†	3·42	13·9	170
White clover roots (living and dead; average of 10 samples over 12 months)	50·2	3·77	13·3	484
Red clover (living and dead ; 5 samples over 6 months)	48·4	2·79	17·5	,,
Lucerne (living and dead; 10 samples over 12 months)	47·4	2·47	20·3	,,
Roots of various legume species	—	1·50–2·70	—	422

* Calculated from analysis of major constituents
† Values for macro-organic matter
— Value not reported

soil after 4 years between grass/clover plots either grazed normally or grazed without return of excreta. Watson and Lapins (476) also found that the rate of N accumulation (72 lb/acre per year) was not significantly affected by whether the grass/clover herbage was (a) grazed, (b) returned as dried plant material, or (c) cut and removed. It is probable that, in these experiments, the amount of N fixed by clover was reduced owing to increased competition from grass stimulated by the return of N in excreta. Furthermore, the N in animal excreta is susceptible to various losses (see Chapter 6) and, of course, to uptake by grass. These factors probably account for the fact that animal excreta have only a small effect on soil N contents.

The rate of application of fertilizer N generally appears to have little effect on the rate of increase in soil N content under grass swards. This is consistent with the finding that although fertilizer N may increase the N content of grass roots (see Table 5), it often decreases root weights (see pp. 63–5) and depresses any clover present.

With grazed grass/clover swards, Clement and Williams (100) found that the application of fertilizer N at rates up to 280 lb for 3 consecutive years had no significant effect on soil N content. With cut swards, soil N contents after 3 years of treatment were lower with high N than where no N was applied. Wolton (509) found that the application of 52 lb N/acre to a grass/clover sward resulted in a lower soil N content than where no fertilizer was applied. Higher applications of N (174 and 312 lb) produced soil N contents similar to those for the unfertilized grass/clover sward. In both these experiments, N fixation by the clover would, of course, have been depressed by the fertilizer N.

With all-grass swards there are a few reports of fertilizer N increasing soil N contents, especially with soils low in OM. In an experiment with subsoil in which the herbage was cut and removed, Sears *et al.* (415) found that urea applied to grass at 562 lb N/acre per year increased soil N over 6 years only slightly less than did the inclusion of clover in the sward. With veld soils in South Africa, applications of fertilizer N to grass cut for hay produced some increase in soil N content (449, 450). Increases in soil OM and N contents resulting from the application of fertilizer N have also been reported in semi-arid soils under cut grass in Nebraska (326) in which initial N contents ranged from 0·072 to 0·185% in the top 6 in. Widdowson *et al.* (491) reported soil N contents of 0·221% and 0·228% from plots that had received no N and a total of 470 lb N/acre, respectively, over a 3-year period. This difference, equivalent to approximately 140 lb N/acre to a 6-in. depth, is unexpectedly high, especially in the light of the previous report (490) by these authors that the average recovery of fertilizer N by the grass was 78%.

An additional reason for the relatively small effect of fertilizer N on the accumulation of organic N in many soils already containing moderate amounts appears to be that it stimulates both immobilization and mineralization. Experiments with ^{15}N (61, 253, 425) have demonstrated that there is appreciable incorporation of added inorganic N, particularly ammonium N, into organic forms. The greater incorporation of ammonium than nitrate N apparently reflects a preference of the soil micro-organisms (58). In contrast to the immobilization of fertilizer N in organic forms, there is evidence that the addition of fertilizer N increases the rate of mineralization of existing organic soil N (an effect known as priming), although the mechanism involved remains uncertain (57). The net effect of fertilizer N is influenced by such factors as the form of N added, the C:N ratio of the OM in the soil, and the level of activity of the soil micro-organisms (58).

A further source of N likely to be significant only in the absence of fertilizer N, is that provided by non-symbiotic fixation (see Chapter 7). This source, together with combined N derived from the atmosphere (see Chapter 13), presumably accounts for any increase in soil N not contributed by legumes or fertilizer.

Factors influencing the rate of accumulation of nitrogen in soil organic matter

The higher levels of OM and N in grassland than in arable soils are thought to result from the greater quantities of plant residues reaching grassland soils, combined with a slower rate of decomposition (208). The latter is often attributed to a lower degree of aeration in grassland soils, and Woldendorp (506) has shown that the respiration of living roots appreciably reduces soil oxygen levels. The high root density in grassland compared with most arable soils would therefore contribute to a slower rate of oxidation of OM.

The actual rate of N accumulation in grassland soils, as in arable soils, is also influenced by the composition of the OM supplied. Contents of C, N, P and S have all been reported to govern the rate of accumulation in certain situations.

Woldendorp *et al.* (508), working in the Netherlands where legumes are not important sward constituents, considered that most grassland soils receive a surplus of C-rich material, while Clement and Williams (100) have suggested that the accumulation of soil N may well depend on the supply of organic C. It is well known that the decomposition of organic materials with a high C:N ratio is liable to result in the immobilization of N, at least temporarily, whereas materials with a low C:N ratio will release mineral N to the soil. Immobilization of N in soil OM is likely to occur when the C:N ratio of the material undergoing decomposition is higher than about 30:1 (173), and detectable mineralization when the ratio is less than 20 or 25:1 (208). The balance between immobilization and mineralization is also influenced by the time-period being considered, since materials with C:N ratios around 30:1 are likely to immobilize mineral N initially, but to release a proportion of this after some weeks as decomposition proceeds.

Some data on C and N contents and C:N ratios in the roots of grasses and legumes are summarized in Table 5. Ratios in the roots of grasses are generally greater than 30:1, whereas in legume roots ratios are generally less than 30:1 and are often below 20:1. Soil N contents have been reported to increase more under grass/clover than under all-grass swards (100, 346, 414, 425), indicating that N fixation by legumes increases soil N more effectively than does either fertilizer N applied to all-grass swards, or non-symbiotic N fixation relying on carbonaceous grass root residues as a source of energy.

The accumulation of C and N under grassland swards may be limited by soil deficiencies of P and/or S which restrict the growth of legumes, and consequently symbiotic fixation of N (469). In a survey of improved pastures in Australia, Donald and Williams (145) found that, on average, the accumulation of soil N to a depth of 4 in. was increased by 85 lb/acre per cwt of superphosphate used. An increase in the rate of accumulation of soil N in response to P application has also been reported by Henzell *et al.* (220). However, Theron (449) reported that although P increased the average yield of veld pasture by 26%, it had no effect on the accumulation of soil C or N. The application of fertilizer S was reported to increase the rate of accumulation of OM under pasture by Williams and Donald (496).

There is a tendency for the N content of soils to show a positive correlation with clay content (436), an effect probably due to the formation of clay-organic complexes and the poor degree of aeration in most clay soils.

The investigations summarized in this chapter indicate that when land previously cultivated is sown to a grass or grass/legume sward, the N content of the soil normally increases at a rate of 50–150 lb N/acre per year. The rate of accumulation is increased by climatic factors favouring high growth rate, by the presence of a legume, by grazing, and by plentiful supplies of nutrients other than N.

CHAPTER 9

RELEASE OF MINERAL NITROGEN FROM PLANT AND ANIMAL RESIDUES AND SOIL ORGANIC MATTER

The decomposition of organic materials results ultimately in a net release of mineral N, even if there is initially a temporary immobilization of N. The net release over a given period, measured either by soil analysis or by plant uptake, represents the difference between the total values for mineralization and immobilization, together with the effects of gains to and losses from the soil. Because of the heterogeneity of field soils, mineralization and immobilization proceed simultaneously.

Both processes are largely the result of the activities of the soil micro-organisms. The production of ammonia from organic materials, known as ammonification, proceeds most rapidly in warm, moist soils but occurs slowly even at 5°C. The ammonia is retained temporarily in the ammonium form on cation-exchange sites in the soil. Only a small proportion normally occurs in the soil solution, and this is subject to uptake by plant roots and to nitrification (see Chapter 10).

The rate of N release depends largely on the C:N ratio of the OM, although other factors may influence the process (see pp. 35–7). Roots and dead herbage derived from grass usually have wide C:N ratios (unless high rates of fertilizer N have been recently applied) and little, if any, N is likely to be mineralized in the early stages of decomposition. Clover roots and herbage, on the other hand, have narrower C:N ratios and release of N proceeds fairly rapidly (see Chapter 4). With regard to animal excreta, the N in faeces becomes available only slowly. The N in urine is mineralized rapidly but is subject to various losses described on pp. 23–4.

There is some evidence (99, 100) that the C : N ratios of soils under permanent grass are higher than those under short-term leys and that the latter have higher C:N ratios than arable soils. This difference is probably due at least in part to the higher proportions of undecomposed root material in the grassland soils, especially when the samples analysed include all living root material. However, the release of mineral N from permanent pasture soil on incubation suggested that the readily decomposable fraction of its OM was not deficient in N (100).

In practice, the annual release of mineral N from the soil can be approximately assessed from the amount of N harvested per unit area in a succession of cuts of grass herbage from swards not receiving fertilizer or clover N. The values obtained will include contributions from atmospheric nitrate and ammonia and from non-symbiotic fixation, but will not include any increase of N during the season in the roots and stubble.

On the basis of yields of N in both grass and cereals as reported in the literature, Walker *et al.* (473) calculated that each 0·1% N in the top 6–9 in. of soil released, on average, 25 lb of mineral N/acre per year, of which 17 lb was harvested. These figures respectively represent approximately 1·25% and 0·85% of the total N content of the soil. Brockman (63) found that herbage N yields from all-grass swards given no fertilizer N ranged from 9 to >80 lb/acre per year at 38 sites in various parts of Great Britain.

It has been shown that the uptake of soil N by grass swards is influenced by their age and botanical composition. Thus, in an experiment at the Hannah Dairy Research Institute, Scotland, on a site that had been under grass for 6 of the previous 7 years, and was ploughed and resown, yields of N in the herbage from a newly established grass sward receiving no fertilizer N were 59, 37 and 29 lb/acre in 3 successive years (391). At Hurley, N yields during 3 consecutive years from two cocksfoot swards were 38, 15 and 20 lb and 32, 25 and 22 lb/acre, respectively (107), and for the 3 years of another experiment with several types of sward, 24, 14 and 17 lb/acre (115, 116). A similar effect was noted by Holliday and Wilman (229). The relatively high N yields obtained in the first years of all these experiments are attributable in part to the increased decomposition of soil OM caused by cultivation. Furthermore, the readily mineralizable N is likely to be taken up mainly in the first year when fresh root residues which might immobilize N would be present only in small amounts.

Species differences in soil N uptake were substantial in the experiments of Cowling and Lockyer (115, 116). Highest average values were shown by S37 cocksfoot (23·4 lb/acre) and S48 timothy (22·4 lb), followed by S24 perennial ryegrass (19·6 lb), Irish perennial ryegrass (19·1 lb) and S215 meadow fescue (18·6 lb); uptake by *Agrostis* sp. (13·3 lb) and S23 perennial ryegrass (11·4 lb) was much lower. A very low uptake of soil N by S23 perennial ryegrass compared with S51 timothy and S215 meadow fescue was also noted by Davies *et al.* (128). However, in an investigation reported by Widdowson *et al.* (490), cocksfoot produced a lower yield of herbage N (average 26 lb/acre per year) than perennial ryegrass (32 lb), timothy (33 lb) and meadow fescue (34 lb).

With permanent grass, and with swards sown on old grassland soils in which the soil OM content had reached equilibrium, quantities of N released may be appreciably greater than those given above. MacFarlan (310) reported that an old permanent pasture consisting mainly of *Agrostis* and fescue with very little white clover, produced in each of two years about 3500 lb herbage DM containing 75 lb N/acre. At the North Wyke Experiment Station, Devon, the average soil N contribution to grass, 2-3 years after ploughing an old pasture and resowing, amounted to about 50 lb/acre per year, while at an old permanent pasture site nearby, the herbage N yield from grazed plots, sprayed in autumn to remove clover, amounted to 187 and 169 lb N/acre without fertilizer N in the two following years (419). Using the plots of a trial in which permanent pasture had been treated with MCPA (216) to eliminate clover, Clement and Cowling (99) found that the N yield in the herbage was 164 lb/acre without fertilizer N, and 261 lb with 104 lb fertilizer N. This annual uptake represented 1·7% of the total soil N (0·45%) in the top 6 in.

A somewhat different method for assessing the importance of soil N was adopted by Hoogerkamp (238, 239), who applied 4 cultivations, each combined with 0, 100, 200 and 300 kg N/ha to permanent pasture plots on a riverine clay soil : (a) no cultivation, (b) deep-digging to 20 cm with the 10– to 20–cm soil layer replaced by topsoil, (c) rotovating to 10 cm, (d) deep-digging to 20 cm with the top 10 cm replaced by subsoil. Plots receiving treatments (b), (c) and (d) were resown with a mixed grass sward. The fertilizer N applications produced similar increases in herbage yields on the 4 cultivation treatments, but (b) produced the highest and (d) the lowest yields. From the N contents of the herbage, the low yields of (d) appeared to be caused by N deficiency, an extra 100 kg N/ha per year being needed to raise the yields of (d) to the level of (a) and (c). Although it is impossible to be certain that the low yield of (d) was due entirely to deficiency of N, the result is consistent with those quoted above.

The release of mineral N from the soil OM sometimes appears to increase following the addition of fertilizer N. According to Broadbent (56), the addition of labelled NH_4- or NO_3-N to a soil increases the mineralization of non-labelled N almost immediately and the net release of N may exceed the quantity added. An increase was also found by Chabannes *et al.* (96) and by Maass (300, 301), who reported that nitrate had a much greater effect than ammonium. However, there is some evidence that mineralization is reduced by the presence of living gramineous plant roots (348, 448), and possibly by some product of the mineralization process (290). The inhibition caused by living roots may thus restrict the supply of soil N to established swards.

Although herbage-N yields from all-grass plots receiving no fertilizer N have varied from about 10 to over 100 lb N/acre per year, depending, to some extent, on the total quantity of N present in the soil and the age of the sward, there is at present no reliable analytical method of assessing the quantity of N likely to be released from the soil to a grass sward (see Chapter 27).

CHAPTER 10

NITRIFICATION, AND IMMOBILIZATION IN NON-EXCHANGEABLE FORM, OF AMMONIUM NITROGEN

The term nitrification is usually taken to mean the conversion of ammonium to nitrate in soils, although Alexander (8) has suggested that all biological oxidations of organic and inorganic N compounds should be included. The review by Alexander includes an assessment of the importance of the various organisms and reactions involved in nitrification.

Grassland soils generally contain inorganic N in amounts ranging from a trace to 5 ppm. (see review by Harmsen and Kolenbrander, 208). This low concentration has caused a number of workers to suggest that nitrification may be inhibited in some way in grassland soils (see 400, 431). The main evidence for this view is that the ratio of ammonium N to nitrate N is generally higher in grassland than in arable soils, and that nitrate content is often extremely low. However, a low content of nitrate is to be expected in grassland soils, since grasses have a high capacity for absorbing N. Also, a relatively high ratio of ammonium N to nitrate N is not necessarily evidence of inhibition of nitrification, since such a ratio could be the consequence of denitrification. Chase *et al.* (97) found no evidence that the presence of grass inhibited nitrification.

The hypothesis put forward by Theron (447) and widely referred to in the subsequent literature, that living plant roots inhibit nitrification by secreting exudates with a bacteriostatic effect on nitrifying organisms, has subsequently been withdrawn (448). His more recent view is that nitrification can take place freely in soil under grass if ammonium N is available, but that the presence of plant roots may retard the release of ammonia from soil organic N.

Both Soulides and Clark (431) and Robinson (400) have pointed out that in some grassland soils, nitrification may be inhibited by acid conditions, especially at pH values below about 5·5.

Robinson also found that in a New Zealand grassland soil, the population of nitrifying bacteria was very small, apparently due to the small amounts of ammonium N present. This would not, of course, apply to soils receiving regular dressings of ammonium fertilizers, and one would expect the ammonium in Nitro-chalk to be nitrified readily. Other ammonium fertilizers may alter the soil pH sufficiently to retard nitrification. Thus, regular applications of ammonium sulphate, unless accompanied by lime, can induce marked acidity. In contrast, the injection of anhydrous ammonia can produce a pH of 9·5 at the site of injection which, together with the high concentration of ammonia present, can also inhibit nitrification (8, 358).

Other factors which influence nitrification are temperature (the optimum being about 30–35°C) and the supply of oxygen and moisture, both of which are required by the nitrifying bacteria (8).

Assessment of the extent to which ammonium N, urea or organic fertilizer N is nitrified is made difficult by the fact that the resulting nitrate may be lost from the soil by plant uptake, denitrification and leaching. However, an indication can be obtained from measurements of soil nitrate levels following the application of fertilizer. Thus, it can be calculated from the results given by McAllister (304) that of 80 lb N/acre applied on 14 March to Italian ryegrass either as ammonium sulphate or as urea, just over 50% was present on 8 April as nitrate in the top 4 in. of soil (assuming 1 acre of soil to a depth of 4 in. weighs 1,332,000 lb). Taking into account that some uptake of nitrate by the grass would have occurred, this result suggests that nitrification was fairly complete within 25 days. Applications made later in the season had much less marked effects on soil nitrate levels, probably owing to more rapid uptake by the grass.

These results also show that appreciable levels of nitrate can occur in grassland soils when the supply of N and nitrification exceed the rate of uptake. This can occur in the absence of fertilizer application in certain conditions. Thus Butler (81) has reported that, in New Zealand, the nitrate content of the soil of pastures of high fertility can increase to high levels when light, warm rains follow prolonged dry periods. Similar accumulations of nitrate in pasture soils during dry periods when grass growth is slow have also been reported from Australia (423).

Most of the ammonium N in soils occurs on cation-exchange sites from which it can be readily displaced by other cations. However, some soils at least have the ability to retain ammonium ions, presumably within the lattice of clay minerals, in such a way that they are not exchangeable with other cations under laboratory conditions. Despite this retention, it is by no means certain that such ammonium ions are completely unavailable to crops (357). Legg and Allison (287), for example, found that ammonium applied to soils capable of fixing ammonium under laboratory conditions was readily taken up by Sudan grass. Freney (165) has even suggested that ammonium fixation is largely a laboratory phenomenon caused by the use of extractants which either contain K and so trap exchangeable ammonium, or (like HF) decompose organic N compounds.

Fixation is greatest in soils which have illite and vermiculite as their major clay minerals, and in soils subjected to drying and heating (14).

Although a greater quantity of fixed ammonium was found by Williams and Clement (497) in a soil under permanent grassland than in a similar soil under arable cultivation, the amount was small in relation to the difference in total soil N, values being 80 ppm. ammonium N for the permanent pasture and 48 ppm. for the arable soil, or the equivalent of 160 and 96 lb N/acre to a depth of 6 in., respectively.

CHAPTER 11

LOSS OF NITROGEN BY LEACHING

Loss of N through leaching from grass and grass/clover swards is generally considered to be small on most soil types. This is an expected consequence of the marked capacity of grass to absorb N. However, there is evidence that losses can be appreciable from very heavily fertilized or intensively grazed swards and from pure clover swards.

Accurate values for leaching losses are difficult to obtain and the published data are scanty. Most lysimeter assessments are of doubtful validity since soil cracks may allow rapid drainage, lack of suction at the base may result in anaerobic conditions and increased denitrification, and soil disturbance with ' filled in ' lysimeters may increase mineralization of N.

Alternative means of estimating the extent of leaching include measuring nitrate N and ammonium N at various depths in the soil profile, and analysis of the drainage water from field drains.

From analyses of soil from two depths, Cunningham and Cooke (119), working with a heavy soil at Rothamsted, found no evidence of leaching during the summer when 112 lb N/acre had been applied on 20 April to the seedbed before sowing ryegrass. Even under bare soil, leaching of N from calcium nitrate, ammonium sulphate or urea was low, and only prolonged rainfall that caused drainage through 20 in. of soil caused leaching of nitrate from the top 9 in. Gasser (172) subsequently found that even for a light soil with little crumb structure, there was no appreciable leaching of N from various fertilizer forms applied at 100 lb N/acre to a ryegrass seedbed in April. Some leaching did occur through soil on uncropped plots.

On the basis of percentage recovery values of applied N in herbage and the distribution of soluble N at depths down to 100 cm in the soil profile, van Burg (76), working in the

Netherlands, concluded that substantial amounts were lost by leaching from 200 kg N/ha applied as ammonium nitrate-limestone to grass swards at various dates between late October and early February.

Support for the view that there is normally little leaching of N from grass swards was provided by a lysimeter study at Hurley (185, 187). Even when attempts were made to induce leaching from grass on a free-draining sandy clay loam overlying chalk by applying 70 lb N/acre in each of 5 consecutive months in winter, only about 10–15% of the applied N was lost in this way. However, anaerobic conditions may have developed at the base of the lysimeters and caused denitrification of nitrate that would otherwise have been leached. On the other hand, lysimeter experiments at Great House Experimental Husbandry Farm, Lancashire, where the average annual rainfall is 60 in. compared with 26 in. at Hurley, showed considerable leaching of fertilizer N. When 224 lb N/acre was applied as ammonium nitrate in 5 equal dressings during the summer to lysimeters inserted into old permanent pasture, average annual losses of nitrate N were 60, 24 and 90 lb in 3 successive years. Without fertilizer, more N was deposited in rainfall than was lost by drainage during the first two years, but there was a net loss of 5 lb/acre in the third year (191).

Losses through leaching from soil under grass in lysimeters have also been reported from the Netherlands. On a sandy soil, losses of ammonium nitrate applied at rates up to 90 kg N/ha were generally small in spring and summer, but there was considerable leaching of N applied in autumn and winter. Even during the growing season, very heavy rain immediately after the application of fertilizer leached 20–35% of the applied N (508). Harmsen and Kolenbrander (208) agree that losses of N by leaching can be appreciable in the Netherlands and state that the loss increases with age of sward. The widespread sandy soils and high average use of fertilizer N in the Netherlands do, of course, favour leaching.

Where leaching of fertilizer N occurs, its extent is influenced by the form applied, since nitrate N leaches much more readily than ammonium N. Urea leaches almost as readily as nitrate, but is normally converted rapidly to ammonium (232), which is in turn converted to nitrate (see Chapter 10).

In New Zealand, where most swards are intensively grazed but receive little or no fertilizer N, high concentrations of nitrate have been found in drainage water. Walker (468) has attributed this nitrate to leaching from urine patches.

Leaching of N where clover is grown alone is considerably greater than from grass or grass/clover swards. In an investigation with carefully prepared lysimeters in S. England, Low and Armitage (295) found that the following average quantities of N were leached over a 3-year period : fallow, 108 lb/acre per year; clover, 27 lb/acre; grass, 2·4 lb/acre. In the following year when the clover was tending to die out, the equivalent of 54 lb N/acre was leached from the clover lysimeters. No fertilizer N was applied during the investigation. Karraker *et al.* (264) also found that leaching of N was considerably greater from lysimeters in which white or red clover was grown alone than when grass was also present. However, much of the leaching occurred in those winters when most of the plants were dead, and when clover growth was good the loss through leaching was less than 10 lb/acre per year.

CHAPTER 12

GASEOUS LOSSES OF NITROGEN

Gaseous losses of N from soils may take the form of ammonia, molecular nitrogen (N_2), nitrous oxide (N_2O), and possibly nitric oxide (NO). Ammonia may be released from ammonium salts when these are applied to calcareous soils, particularly under warm conditions, and also from urea. Losses of ammonia from animal excreta are discussed in Chapter 6 and from fertilizers in Chapter 22.

Losses of N as N_2 and N_2O are generally considered to be due mainly to denitrification, i.e. the microbiological reduction of nitrate. However, chemical reactions resulting in N_2 or nitrogen oxides may also occur in soils and are discussed below. In general, gaseous losses appear to be important only when rather large concentrations of soluble N occur in the soil following the addition of fertilizer or animal excreta.

The difficulties of estimating all gaseous losses of N in the field are considerable (324) and no quantitative determinations have been made under field conditions. Consequently, the main evidence for denitrification in the field has been derived from balance experiments in which recoveries of fertilizer N in herbage, soil and sometimes leachate have amounted to appreciably less than 100%. Such values, obtained by difference, are of course subject to the cumulative errors arising from the other determinations involved (323, 324), and to underestimation due to undetermined gains.

Since it involves the microbiological removal of oxygen from the nitrate ion, denitrification occurs to an appreciable extent only in conditions of low oxygen partial pressure which, in soils, result mainly from high moisture content, the presence of decomposing organic matter, and root respiration.

Numerous experiments have shown that denitrification losses are greater when soil conditions are, at least temporarily, anaerobic (20, 55, 356, 507). Patrick and Wyatt (376) reported that successive cycles of submergence and drying resulted in the loss of 15–20% of the total N from a soil containing 986 ppm. N. Most of the loss occurred during the first 2–3 cycles. Presumably mineralization and nitrification predominated during the relatively dry, and denitrification during the relatively wet, phase. No loss of N was measured from soil kept at optimum moisture content. However, Dilz and Woldendorp (141) found no difference in the extent of gaseous loss of N between soils maintained at either 80 or 90% of their water-holding capacity.

Greenwood (195) has suggested that the occurrence of denitrification under conditions of apparently good aeration may be explained by the existence of anaerobic zones in the interior of soil crumbs. The need for suitable H-donors from soil OM, as well as conditions of oxygen deficiency, has been stressed by Woldendorp *et al.* (508), Broadbent and Clark (59) and Woldendorp (507), who quoted results showing a relationship between denitrification and the OM content of soils. The results of Ekpete and Cornfield (155) indicated that when OM was applied to a soil together with nitrate, heavy denitrification losses could occur even at low soil moisture levels. Simpson and Freney (425) reported a loss of about 30% from sodium nitrate added to a soil containing about 7·5% OM in both the presence and absence of newly sown ryegrass; losses from two soils containing less than 4·2% OM were negligible.

The importance of root respiration in lowering soil oxygen levels was indicated by Woldendorp (506), who found that the presence of living grass roots in a soil increased denitrification as compared with dead roots. Oxygen consumption by living roots was 28 times greater than that of dead roots, two-thirds of the consumption being attributable to root respiration and the remainder to the rhizosphere microflora. In addition to decreasing soil oxygen content, Woldendorp suggested that living roots may excrete organic compounds which act as H-donors. Gasser *et al.* (175) also found greater gaseous losses from nitrate added to soils on which ryegrass was grown than when the soils were incubated.

Denitrification is slow in acid soils, the optimum pH being 7·0–7·5 (208). The optimum temperature is high but appears to cover a wide range (approximately 28–60°C) (507). Little denitrification occurs at temperatures below about 10°C (208, 507) and losses of N by this process are therefore unlikely to be important during winter in temperate regions.

The evidence summarized by Woldendorp *et al.* (508) indicates that, except in winter, conditions are often very favourable for denitrification in grassland soils. Root density is high, thus reducing soil oxygen content to a low level and at the same time providing a supply of H-donors from root excretions. Furthermore, the number of denitrifying organisms in the rhizosphere is much higher than in non-rhizosphere soil and, particularly after the application of fertilizer N, there is a substantial flow of nitrate through the rhizosphere zone.

Losses of fertilizer N thought to have been due to denitrification have often amounted to 20–30%. Using cores of permanent grass swards from which leaching was prevented, Dilz and Woldendorp (141) found that losses of nitrate applied as KNO_3 at a rate equivalent to 225 lb N/acre amounted to 16–31% for a clay soil, 11–25% for a sandy soil and 19–40% for a peat soil. In a pot experiment with grass and grass/clover mixtures grown in a subsoil low in total N, Walker *et al.* (471) found losses of 25–35% of the N supplied as labelled KNO_3 or $(NH_4)_2SO_4$ at rates up to 200 ppm. In both these experiments the losses were apparently gaseous, since the N contained in roots was determined and leaching prevented.

In addition to microbiological denitrification, it is possible for loss of N to occur through chemical reaction. Allison (12) considers that wherever appreciable quantities of ammonium and nitrite ions are present simultaneously, especially in acid soils, some loss of gaseous N is a strong possibility. The chief reaction is thought to be the formation and decomposition of ammonium nitrite :

$$NH_3 + HNO_2 \rightarrow NH_4NO_2 \rightarrow N_2 + 2H_2O$$

The self-decomposition of HNO_2 may also be important in acid soils of low exchange capacity (11):

$$3HNO_2 \rightarrow 2NO + HNO_3 + H_2O$$

A third possible reaction (507) is that between nitrous acid and organic compounds which have an amino group in the α position with respect to a carboxyl group :

$$RNH_2 + HNO_2 \rightarrow ROH + H_2O + N_2$$

The practical importance of these reactions and the effects upon them of metal ions and other soil constituents require further investigation (507). These chemical reactions are unlikely to be appreciably reduced by low temperatures unless the ground is frozen.

Non-biological losses of this type may have occurred in the experiment carried out by Martin *et al.* (323) in which Rhodes grass was grown in pots in a light-textured soil of pH 5·6. Of the N added as $^{15}NH_4NO_3$, 93·36% of the total and 94·0% of the labelled N was recovered in the soil/plant system. The fact that these values did not differ significantly indicated that the loss fell equally on the labelled ammonium N and on either the nitrate N or the native soil N, and consequently that biological denitrification of nitrate did not offer a satisfactory explanation.

It is impossible to give any reliable estimate of the importance of denitrification and similar chemical reactions in grassland soils, since no field assessments have been made. Laboratory and pot investigations have shown that environmental factors such as soil moisture status, pH, temperature and rate of OM decomposition have a marked effect on gaseous losses.

CHAPTER 13

ADDITIONS OF COMBINED NITROGEN FROM THE ATMOSPHERE

Small quantities of nitrate N and ammonium N occur in the atmosphere and may reach the soil in precipitation or by direct adsorption. Accurate assessments are difficult to make owing to the low concentrations involved. The quantity of N carried down in precipitation appears to be greatly influenced by proximity to industrial sources. In a review of data from a large number of sites in Europe, Eriksson (158) reported an average value of 7·2 lb/acre per year, with individual values ranging from 0·7 to 19·6 lb. Data on the composition of rain-water over the period 1959–64 from 10 widely distributed sites in Great Britain, chosen to avoid local industrial pollution, have been reported by Stevenson (435). Calculations based on the median concentrations of nitrate N and ammonium N, and rain-fall data for the various sites, indicate that accessions of N ranged from 2·7 lb/acre per year at Lerwick in the Orkneys, to 18·8 lb at Newton Abbot, Devonshire. 8 of the 10 sites produced values in the range 2·7–5·9 lb/acre. At Jealott's Hill, Berkshire, the average annual deposition of N (including organic N) in rainfall during 1953–6 amounted to 17·5 lb/acre (295).

In addition to nitrate and ammonium, a small amount of organic N is deposited as bacterial cells and dust. Allison (12) considers that this generally amounts to less than 1·5 lb N/acre per year, but Low and Armitage (295) have reported a value of 6·3 lb/acre, approximately one-third of the total N.

There is conflicting evidence, summarized by Allison (12), as to whether direct adsorption of ammonia by soil is of any significance in crop nutrition. Appreciable adsorption of ammonia from the atmosphere by bare soil was reported by Malo and Purvis (320) at Rutgers, New Jersey, daily rates varying from 0·05 to 0·20 lb N/acre. The average annual total was equivalent to 44 lb N/acre, with lower amounts in sandy soils. Evidence that 20–60 lb N/acre per year might well be added to the soil by the adsorption of atmospheric ammonia was obtained, also in New Jersey, by Hannawalt (205).

These results suggest that some of the gains previously attributed to non-symbiotic fixation may have been due to adsorption of N from this source. Soils with a complete vegetation cover, such as established grassland, would be expected to adsorb less ammonia than bare soils. However, the possibility of direct absorption by the herbage remains, though this does not appear to have been so far investigated.

CHAPTER 14

NITROGEN BALANCE IN GRASSLAND

It is impossible to assess accurately in a field situation all the gains and losses of N occurring during the course of a growing season. Methods for measuring some of the components of gain and loss necessarily impose artificial conditions on the system, and for grazed swards, the position is further complicated by the fact that some of the N in the system may circulate from soil to livestock and back to the herbage several times during a season.

Accurate assessments can, of course, be made of N inputs in fertilizer, of N removals in herbage and of the N in the unharvested portion of the sward. Fairly precise estimates can also be made of symbiotic fixation in grass/legume swards. Leaching losses can be measured by means of lysimeters, though the values obtained may differ from those occurring in undisturbed soil. Changes in soil N content can also be assessed, and with increasing accuracy, as the quantities involved increase with sward age. Non-symbiotic fixation, accessions from the atmosphere and denitrification are impossible to measure individually under normal field conditions, even if labelled N is used; even under laboratory conditions these components are difficult to measure.

In order to obtain a moderately complete N balance an estimate must be made of leaching losses, since though these are generally small from grassland (except on sandy soils), nitrate is a very soluble soil constituent. Applied N not accounted for in plant, soil or leachate can then be attributed to denitrification or other gaseous losses. A number of lysimeter investigations with grass swards have been carried out, but because of uncertainty as to the validity of their results and the cost and labour involved, interest in such studies has declined in recent years. Most lysimeter work has been carried out in the USA and the results have been summarized by Allison

(10, 13) and Martin and Skyring (324). Results of the experiments carried out with timothy at Ithaca, New York State, are shown in Table 6.

TABLE 6

Annual N balance in lysimeter experiments with timothy (Bizzell (41), summarized by Allison (10))

(Values are averages for an 8-year period)

	lb N/acre/year			
N added as $NaNO_3$	93	124	155	213
N in rain	6	6	6	6
Total input	99	130	161	219
N removed in cut grass	78	98	121	154
N gain in soil	15	19	14	17
N present in leachate	2	1	2	3
Total N accounted for	95	118	137	174
N not accounted for	4	12	24	45
N input not accounted for, %	4	9	15	21

No assessments were made of non-symbiotic N fixation or of direct adsorption of atmospheric ammonia by soil or grass; hence if these gains were significant, the N not accounted for (presumably due to gaseous losses) would have been correspondingly greater.

In the lysimeter experiment at Hurley mentioned on p. 46, a total of 1330 lb N/acre was applied over a 4-year period up to the end of 1962, and no further fertilizer was applied in 1963 or 1964. 500 lb of the applied N was unaccounted for in the autumn of 1963 and only 50 lb was recovered in herbage and drainage water in 1964. Changes in soil N, gains from the atmosphere and non-symbiotic fixation were not assessed in this experiment, but assuming an annual increase in soil N of 50 lb/acre, 10 lb supplied from the atmosphere and no non-symbiotic fixation, there would still be a net loss of 230 lb, amounting to 17% of the total applied.

Other experiments in which N balances were obtained and gaseous losses estimated by difference are described on pp. 50–5.

As pointed out by Woldendorp *et al.* (508), the magnitude of undetermined and (presumably) gaseous losses appears to increase with increasing input levels of fertilizer N. Most lysimeter experiments with high N rates have shown losses of N, whereas some experiments with low rates have shown gains. This indicates that with low rates of application, losses are small and maybe exceeded by gains from non-symbiotic fixation and from the atmosphere. With heavy applications, gains by fixation are probably reduced and losses are apparently greater.

The use of labelled N in lysimeter experiments increases the accuracy of assessment of several of the components in N-balance determinations and eliminates non-symbiotic fixation from calculations of N loss from applied N ; however, this technique does not appear to have been used with grass.

On the basis of the results discussed in Part I of this review, it is possible to state probable values for the quantities of N involved in the various phases of the N cycle under different systems of grassland management, and this is attempted for 4 systems in Figs 4-7. It should be emphasized that most of these values will vary considerably with local conditions and, for the systems receiving fertilizer N, with the factors affecting response

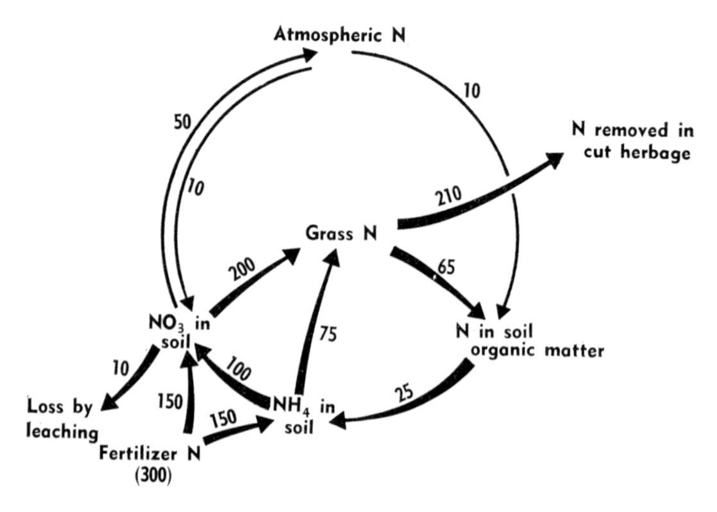

Fig. 4. N transformations in a cut grass sward receiving 300 lb N as ammonium nitrate/acre/year. Numerical values are estimates in lb N/acre/year.

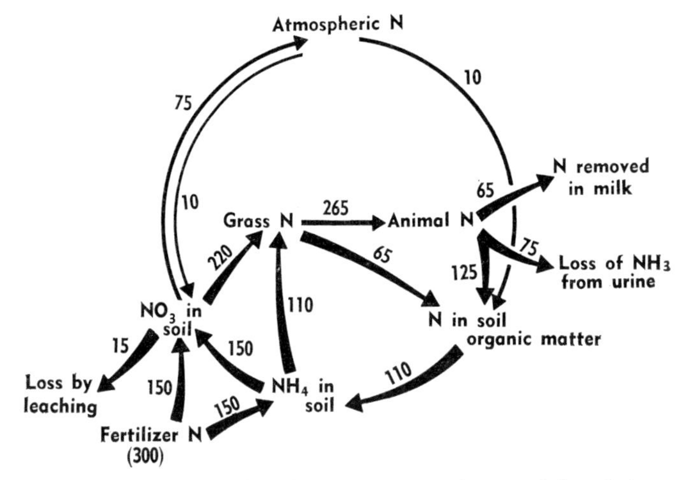

Fig. 5. N transformations in a grass sward grazed by dairy cattle and receiving 300 lb N as ammonium nitrate/acre/year. Numerical values are estimates in lb N/acre/year.

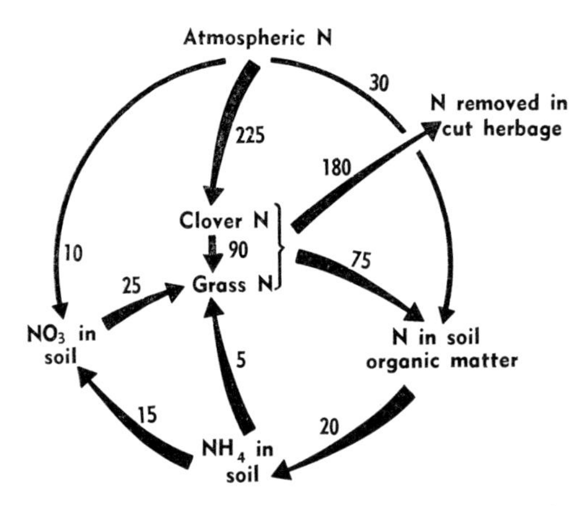

Fig. 6. N transformations in a cut grass/clover sward not receiving fertilizer N. Numerical values are estimates in lb N/acre/year.

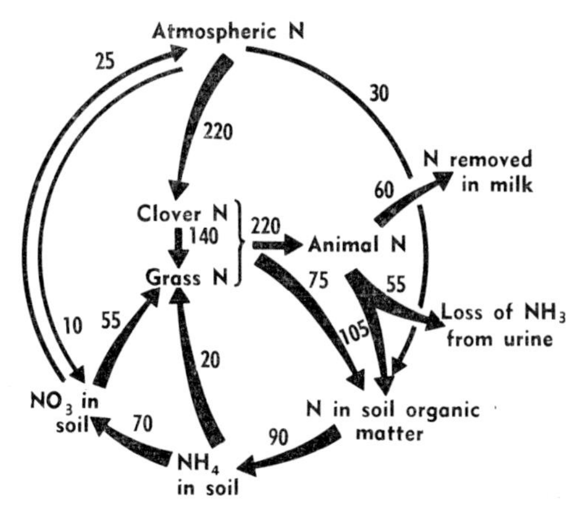

Fig. 7. N transformations in a grass/clover sward grazed by dairy cattle and not receiving fertilizer N. Numerical values are estimates in lb N/acre/year.

discussed in Part II. The values are intended to apply to conditions in southern England.

A somewhat similar analysis of the N economy of grazed grassland in South Africa was made by Davidson (124). From experimental results, he concluded that under his conditions, better use of fertilizer N was achieved by giving a relatively large dressing of N (about 100 lb/acre) in the first year of a 5-year period and about 30 lb N in each of the subsequent years, rather than by dividing the same total amount equally between years.

PART II

THE YIELD RESPONSE OF GRASS SWARDS TO FERTILIZER NITROGEN

CHAPTER 15

EFFECTS OF NITROGEN SUPPLY ON THE MORPHOLOGY AND PHYSIOLOGY OF GRASS PLANTS

N affects herbage yield through its influence on various aspects of the morphology and physiology of the grass plant. Factors such as tiller production, leaf area and root growth are all modified by N supply.

Most studies involving assessments of changes in morphology and physiology have been carried out on individual plants grown singly either in pots or in the field; there is, however, some evidence that the effects of N are different under sward conditions where competition between plants is important.

Effect of nitrogen on the rate of tiller production

With single plants, increasing the supply of N over the range at which there is an increase in herbage yield, increases the number of tillers per plant. Under dense sward conditions, the rate at which tillers are initiated may be increased by increasing N supply, but many of the tillers are likely to be short-lived.

The effect of 3 levels of nitrate N, in combination with 3 levels of P and K, on individual plants of timothy growing in sand culture out-of-doors was studied by Langer (278). The average number of tillers per plant for the 3 levels of N at the highest levels of P and K at the final harvest was 32 for the 6-ppm.-N level, 70 for the 30-ppm. and 141 for the 150-ppm.

level. The 6-ppm. level represented a fairly severe N deficiency. Similar results were obtained in subsequent experiments also carried out with timothy (279).

The influence of N supply on tillering in cocksfoot grown in soil/sand mixtures under glasshouse and growth chamber conditions was examined by Auda *et al.* (23). Under favourable conditions of light and temperature, the application of N at a rate equivalent to 200 lb/acre increased tiller numbers at least 3-fold compared with no N.

In an experiment in which individual grass seedlings were grown in small plots out-of-doors in New Zealand, O'Brien (367) obtained the following tiller numbers at 8 weeks after germination :

N applied, lb/acre	0	6	20	60	120
Perennial ryegrass, tiller no.	11	16	27	67	113
Cocksfoot, tiller no.	3	7	16	33	46

Ryle (405) grew cocksfoot plants under simulated sward conditions on vermiculite in a controlled environment. Plants supplied 3 times per week with solution containing 15 ppm. N had an average of only 2 tillers per plant after 12 weeks, compared with 4-5 tillers for those on solution containing 150 ppm. N.

As pointed out by Langer (280), the effect of N supply on tillering appears to increase with time and is more marked with secondary than with primary tillers. Plants deficient in N stop producing tillers at an early stage.

Under sward conditions the effect of N on tiller number is generally smaller than that reported for single plants, a difference probably due mainly to the effects of competition. Bean (35) found that compared with 17 lb N/acre, 52 lb increased tiller number/unit area by about 65% in a perennial ryegrass sward at about 7 weeks after application; with cocksfoot the effect was much smaller.

With regularly cut swards of Italian ryegrass, D'Aoust (122) found that N and water supply interacted in their effects on tillering. Under low-moisture conditions during the dry part of the season, the number of tillers per unit area was substantially higher with an application rate of 224 lb N/acre per year than with 112 lb N, but the number with 448 lb was similar to that with 224 lb. With irrigation to prevent the development of soil moisture deficits greater than 1·5 in., there was little or no difference between the 112- and 224-lb N treatments, but tiller number with 448 lb was appreciably higher.

In an experiment in which cores cut from a sward of timothy were grown in growth chambers under cool or warm conditions, Smith and Jewiss (428) found that ammonium nitrate applied at the rate of 300 lb N/acre per year, increased tiller number/unit area by up to 50% after several weeks under cool conditions, but that the increase was much less under warm conditions.

Summarizing the results of these experiments, it is clear that increasing the N supply usually increases tiller number, but the extent of the increase is modified by other environmental factors such as plant density, water supply and temperature.

Effect of nitrogen on the rate of leaf production per tiller

Level of N appears to have little effect on the rate of leaf production per tiller, irrespective of whether the plants are grown as spaced plants or in swards.

In Langer's investigation (278), in which a sand culture technique was employed, mean values for leaf number per tiller in timothy for 3 levels of N were as follows :

N, ppm. in solution	6	30	150
Leaves/tiller, average no.	3·10	3·45	3·41

A study with 6 species of grass reported by Ryle (404) also showed only a small, though significant, effect of N on leaf number. Plants grown on vermiculite in a glasshouse during the period 29 December to 4 April had the following mean leaf numbers :

Nitrate-N, ppm. in solution	15	150
Leaves on main shoot, average no.	8·9	9·6

In a subsequent investigation with cocksfoot grown as a simulated sward in a constant environment, plants receiving 150 ppm. N in nutrient solution had slightly fewer leaves per tiller than plants receiving 15 ppm. N (405).

With swards, Bean (35) found little difference between rates of 17·4 and 52·2 lb N/acre per cut in their effect on the mean number of leaves per tiller in S23 perennial ryegrass and S37 cocksfoot; and O'Brien (367) found no difference in the mean number of leaves per tiller of seedling ryegrass and cocksfoot plants receiving either no N or 120 lb N/acre.

Effect of nitrogen on leaf size

Although N supply appears to influence herbage yields mainly through its effect on leaf size, there is a lack of experimental data to confirm this, particularly under field conditions.

For individual plants of timothy grown with adequate P supply, Langer (278) reported the following mean leaf areas for 6-month-old plants :

Nitrate-N, ppm. in solution	6	30	150
Average leaf area, cm^2	0·94	1·37	2·30

In subsequent experiments, Langer (279) assessed total leaf area for groups of 3 plants, but not individual leaf areas. The large differences in total leaf area between N treatments were attributed partly to effects on leaf number (mainly a reflection of tiller numbers) and partly to effects on leaf size.

Ryle (404) found that differences in leaf size in 6 grass species receiving either 15 or 150 ppm. N in the nutrient solution were generally not significant in seedlings until the 9th leaf had developed. In an experiment using a simulated sward of cocksfoot in constant-environment conditions, a level of 150 ppm. N in the nutrient solution applied 3 times per week increased the average area of individual leaves from 8·5 (with 15 ppm. N) to 13·5 cm^2, mainly by increasing leaf length (405).

Working with swards, Bean (35) found that increasing the N application rate from 17·4 to 52·2 lb/acre per cut increased the mean areas of mature leaves in S23 perennial ryegrass and S37 cocksfoot by approximately 50%.

Effect of nitrogen on the longevity and metabolic activity of leaves

The effects of N on the longevity and metabolic activity of leaves appear to be small except at very low levels of supply.

Langer (279) concluded that a decline in net assimilation

rate (increase in dry weight/unit leaf area per unit of time) and leaf area ratio (ratio of leaf area to plant dry weight) of timothy occurred only at very low levels of N supply. The product of net assimilation rate and leaf area ratio, known as relative growth rate, was lower for the 6-ppm.-N treatment than for the 30-ppm. and 150-ppm. treatments up to 12 weeks from the time the treatments were imposed, but not thereafter. Ryle (405), however, found a consistently higher value for net assimilation rate in simulated swards of cocksfoot receiving nutrient solution containing 150 ppm. N than in similar plants receiving 15 ppm. N. Leaf area ratio was also greater with the higher level of N after 3 and 6 weeks of treatment, but not after 9 and 12 weeks. The effect of N on relative growth rate therefore decreased with time. O'Brien (367) also found that the effect of N on relative growth rate declined with time; compared with no added N, 60 lb N/acre increased the relative growth rate 5-fold in the first two weeks, but only 2-fold in the 2nd- to 4th-week period.

The effect of N supply on net assimilation rates of timothy and meadow fescue was found by Lambert (277) to depend on cutting frequency. The two levels of N he used (52 and 156 lb N/acre per year) caused no difference in net assimilation rates in swards cut for hay and cut twice thereafter, but the 156-lb rate gave appreciably higher rates of net assimilation with monthly cuts.

Bean (35) found no difference in the rate of leaf senescence between plants grown with 17·4 or 52·2 lb N/acre per cut.

Effect of nitrogen supply on root growth

It is well established that increasing the supply of N almost always increases the shoot/root ratio. Although root growth is increased by an increase in N supply (but to a lesser extent than shoot growth) under conditions of severe N deficiency, once a moderate supply of N has been attained, further increases often reduce root weight (65, 453). As pointed out by Garwood (169), differences in root weight do not necessarily correspond to differences in root growth, since the rate of decomposition may also be increased. Such an effect was noted by Vose (463) with perennial ryegrass in culture solution ;

roots of plants in solutions containing 84 ppm. N became brown and sloughed off cell wall material, while roots in solutions containing 16·8 or 4·2 ppm. N remained white and healthy. However, root elongation was reduced by fertilizer N in an experiment with grasses grown in soil in a glass-sided box (370), in which the following results were obtained :

N applied, lb/acre	0	300
Bromegrass root elongation, mm/24 h	36·0	28·1
Cocksfoot root elongation, mm/24 h	30·4	22·6

Field observations confirmed this effect. Roots of plants receiving no N had, in August, reached a depth of 32 in., but only 20 in. in plots receiving 600 lb N/acre per year; root weights were also substantially reduced.

Mitchell (334) reported the following root weights for cocksfoot in 6-in. cores to a depth of 24 in. for two levels of ammonium nitrate application in each of 3 years:

	1957		1958		1959	
N applied, lb/acre	0	150	0	150	0	150
Cocksfoot root weight, g	6·17	5·95	8·32	6·58	3·68	2·89

Reductions in grass root weights induced by increases in N supply in field experiments have also been reported by Lambert (276) and Garwood (169). Holt and Fisher (237), however, found no reduction in the root weight under a sward of Coastal Bermudagrass where ammonium nitrate was applied at up to 1600 lb N/acre. In pot experiments, decreases in grass root weights with increasing N supply were obtained by Blackman and Templeman (46), Nielsen and Cunningham (355) and Auda *et al.* (23).

In addition to influencing the overall growth of roots, N supply also affects the morphology and the physiological activity of root systems. Oswalt *et al.* (370) reported a clear increase in root diameter as the rate of N application increased up to 300 lb N/acre, but at the same time a decrease in root number. An increase in the efficiency of uptake of nutrients other than N is indicated by the fact that increasing the supply of N greatly increases the yield of DM and often the percentage

P and K contents in the herbage, without causing a corresponding increase in root weights. Good correlations between N supply, N percentage in the roots and cation exchange capacity of the roots for a number of plants, including cereals and grasses, were reported by McLean *et al.* (313); and Mitchell (334) found with cocksfoot that the absorption of the organic cation methylene blue by root tissue, which is regarded as a measure of root activity, was increased by about 15% by the application of 150 lb N/acre per year.

CHAPTER 16

INFLUENCE OF LIGHT AND TEMPERATURE

Although the growth of grass is known to be very much influenced by both light and temperature, few investigations of their influence on the response of grass to fertilizer N have been carried out.

In field and pot experiments in which *Agrostis tenuis* and *Festuca rubra* received either no N or the equivalent of 37·5–50 lb N/acre every 7 days, Blackman and Templeman (45) found that light intensity greatly influenced N response. Under full natural light intensity during June—October, N application markedly increased herbage yield, whereas at 60% normal light intensity, responses to N were small and at 40% normal light, N application actually reduced DM yield in 3 out of 4 experiments.

In outdoor plot trials with perennial ryegrass, Deinum (130) imposed rates of 25 and 125 kg N/ha in combination with 3 light intensities : (a) natural daylight during mid-September to mid-October, average level 240 cal/cm^2 per day, (b) reduced light intensity equivalent to a dull day in mid-winter, average level 48 cal/cm^2 per day, and (c) increased light intensity equivalent to that of midsummer, average level 547 cal/cm^2 per day. His results showed that the response to N increased with increasing light intensity. However, light intensity had surprisingly little effect on the herbage yields (which were high) obtained with only 25 kg N/ha. Deinum confirmed the effect of light intensity in glasshouse experiments. Rather similar results were obtained by Cunningham and Nielsen (120) with Italian ryegrass grown in soil in pots in midsummer, in which shading to 44% of the normal light intensity caused a much greater reduction in the response to 100 ppm. N than did shading to 68%.

In addition to seasonal variations in cloudiness and in the

elevation of the sun, the amount of light available for photosynthesis depends also on daylength. In a pot experiment in which Wimmera ryegrass plants were grown in controlled-environment conditions, Collins (101) found that their yield response to N at 92 and 184 lb/acre was 30–50% greater in a daylength of approximately 12 hours than in an 8-hour daylength. He suggested that the short daylength of about 8 hours was an important factor restricting the response of grass to N in winter in Victoria, Australia.

The effect of temperature on response to N has not been clearly demonstrated, since the results of field experiments have been complicated by variations in other factors and the pot experiments so far reported appear to have been restricted to temperatures above 10°C.

Field investigations by Blackman (44) indicated that no growth of grass swards occurred, even in the presence of fertilizer N, when the soil temperature at a depth of 4 in. was less than 42°F (5·6°C); that at temperatures between 42° and 47°F (8·3°C) growth was increased by fertilizer N, and that at temperatures above 47°F growth rates were similar with and without fertilizer N. The effect of fertilizer N on herbage yield in spring was therefore greatest in years when the soil temperature rose slowly from 42 to 47°F. Blackman suggested that the lack of effect of fertilizer N on growth rate at temperatures above 47°F was due to the substantial release of mineral N from the soil OM. However, this result is unusual and presumably reflects (a) high soil OM contents in the pasture soils of the experiments and (b) the relatively early harvesting dates, which would have reduced the depletion of mineral N from the soil.

In a pot experiment with Italian ryegrass, Parks and Fisher (374) found that at soil temperatures of 10°, 20° or 30°C there was a higher response to N (applied at a rate equivalent to 100 lb/acre) at 20° than at 10° or 30°C. In a rather similar experiment in which the 3 temperatures investigated were 11°, 19·5° and 28°C, Nielsen and Cunningham (355) found that although yields were higher at 19·5° than 11°, the response to 100 ppm. N was almost identical at these two temperatures, and substantially lower at 28°. In the glasshouse experiments carried out by Deinum (130), although grass yields generally increased somewhat with increasing temperature over the range 10–23°C, there was no effect of temperature on response to applied N.

The results of these experiments suggest that the response of grass swards to fertilizer N may commonly be limited by low light intensity such as may occur in dull weather during summer and, of course, to a greater extent at other times of year. More information is required on the influence of temperature, but it seems likely that this factor is more important than light in restricting grass growth in early spring in cool temperate regions. The effects of light and temperature on response to N are due to their effects on plant growth rather than to any specific effect on N uptake.

CHAPTER 17

INFLUENCE OF CLOVER

In the absence of fertilizer N, yields from grass/clover are higher than from all-grass swards, owing to the ability of clover to fix atmospheric N. The application of fertilizer N reduces clover growth and therefore N fixation (see Chapter 2); responses to fertilizer N, in terms of herbage yield increase per lb N applied, are therefore lower with grass/clover swards.

Despite the practical importance of this topic, relatively few direct comparisons of the response to fertilizer N of grass and grass/clover swards have been reported. Results from 5 such comparisons (64, 107, 184, 391, 420) are, however, included in Fig. 8, together with data on the response of a grass/clover sward (291).

Whereas the average annual response for all-grass swards is about 25 lb DM/lb N, that for grass/clover swards is about 12 lb. Grass/clover swards show greater variation in response than all-grass ones, since the former differ in the proportion of clover present and because clover is more susceptible than grass to adverse growth conditions. When the proportion of clover is small and conditions unfavourable for clover growth, yields and responses will be similar to those of all-grass swards. On the other hand, swards with a large clover component which produce high yields without fertilizer N, may show a negative response to N applications. Thus, Cowling (107) reported that in one year of his trials, the yield of a cocksfoot/white clover sward was decreased even by 210 lb N/acre per year.

From the results shown in Fig. 8 and from other investigations in which grass/white clover swards not receiving fertilizer N have been compared with grass swards receiving a range of N rates (18, 115), it appears that on average in Britain, an all-grass sward requires about 140 lb N/acre per year to achieve the same herbage yield as a grass/clover sward not receiving

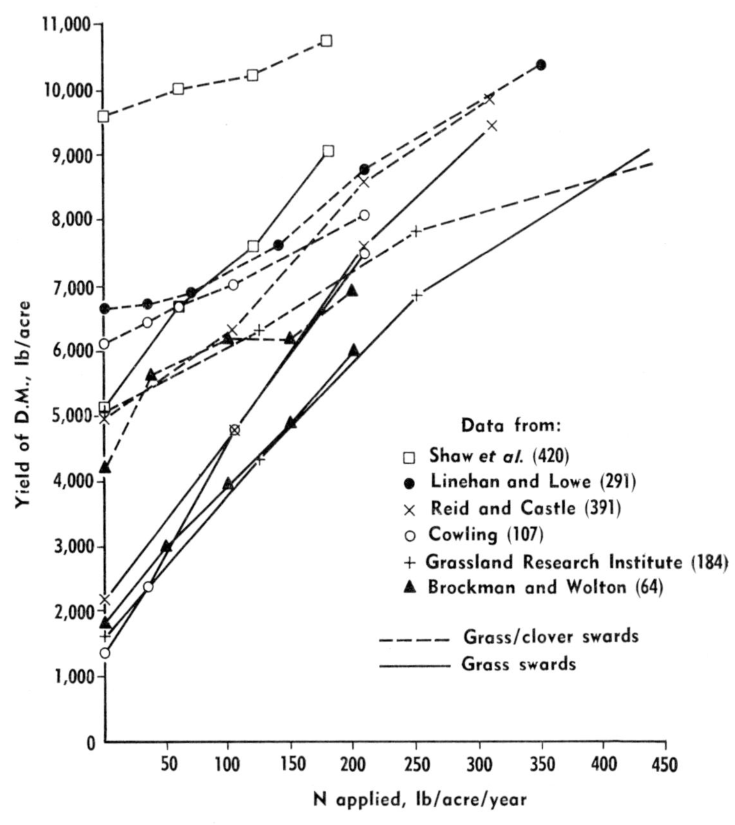

Fig. 8. The responses of grass and grass/clover swards to fertilizer N from several experiments; all cut swards.

fertilizer N. Again, there is considerable variation in this figure from year to year and from site to site, which largely reflects the variation in the yields obtained from grass/clover swards not receiving fertilizer N. Thus, annual yields at Hurley average about 5500 lb DM/acre, but a range of 2900 to 8000 lb has been recorded from individual well-managed swards. Other centres have reported higher yields, e.g. 10,000 lb DM/acre (4-year average) in the Republic of Ireland (351); 9520 lb in the West of England (18); 6900–9200 lb in S.W. Scotland (94). In New Zealand yields of 12,000–15,000 lb

DM/acre from grass/clover swards can be obtained (156, 412), equivalent to those obtained from all-grass swards receiving 350–400 lb N/acre.

When such high yields from grass/clover swards can be relied upon, the response to fertilizer N is so low as to make its general use uneconomic, a situation common in New Zealand (156). Where clover growth is generally poor, the economic value of fertilizer N is higher, especially where maximum production per unit area is important. Thus in the Netherlands, the average use of fertilizer N on grassland is high (148 kg/ha in 1963–4) (372) and clover is often disregarded.

In Britain, the relative economic merits of clover N and fertilizer N and the possibilities of combining the two sources are subjects of continued discussion. The choice is much influenced by the climate of the area in question and by the farm management system employed. With permanent grassland, especially in sheep-grazing systems in hill areas and in the wetter western and northern parts of the country, considerable reliance is placed on clover. In the southern and eastern areas and particularly with temporary grass and where much of the crop is cut for conservation or grazed by dairy cattle, it is increasingly common to employ all-grass swards receiving more than 300 lb fertilizer N/acre per year.

Whether it is generally practicable to combine clover and fertilizer as sources of N is more difficult to resolve. The results shown in Fig. 8 indicate that higher yields are obtained from grass/clover than from all-grass swards at rates of fertilizer N up to 200–350 lb/acre per year, and therefore that it may be economic to include clover in swards even when moderate rates of fertilizer N are to be applied. The combination of grass/clover swards with the use of about 150 lb N/acre per year has been recommended as satisfactory by Brockman and Wolton (64). However, the success of this approach depends on the establishment and maintenance of the clover component of the swards. Although achieved in the experiments reported, the establishment and maintenance of clover introduces additional difficulties into sward management. If, however, the necessary management requirements can be met, the inclusion of clover in a sward intended to receive 100–200 lb N/acre per year should be worthwhile. Castle (91) has suggested that it might be practicable to rely mainly on clover in the early years after sowing and to increase the rate of fertilizer N as the vigour of the clover declines. Other workers consider that, for any one sward, it is best either to depend completely on clover for N supply, managing the sward in such a way as to promote N fixation, or to omit clover and rely on fertilizer N (291, 468, 475).

CHAPTER 18

INFLUENCE OF SPECIES AND VARIETY OF GRASS AND AGE OF SWARD

Differences between cultivated grass species and varieties in response to fertilizer N are usually small and, at high levels of N input, age of sward appears to have little effect on herbage yields.

TABLE 7

Yield responses of various grasses to fertilizer N, lb DM/lb N
(Responses are based on yields from no-N plots)

	Hannah Dairy Res. Inst., Ayrshire			Rothamsted, Herts	Hurley, Berks		North Wyke, Devon
Reference	236		391	490	115		513
Date	1949–52		1960–62	1958–60	1960–62		1959–61
lb N/acre/year	140 or 210	350 or 410	312	375	140 or 174	280 or 348	0 or 75–300
Species and variety							
Ryegrass :							
S23	—	—	24·7	—	24·0	23·5	—
S24	—	—	—	30·8	26·2	23·4	16·7
S101	16	17	—	—	—	—	—
Ayrshire	16	15	—	—	—	—	—
Irish	—	—	—	—	22·4	18·3	—
Cocksfoot:							
S26	29	19	—	—	—	—	—
S37	—	—	22·7	45·6	30·5	25·0	16·7
Scotia	23	18	—	—	—	—	—
Danish	29	20	—	—	—	—	—
Timothy :							
S48	21	17	—	—	29·5	22·6	—
Scotia	18	15	—	36·9	—	—	—
Scots	23	17	—	—	—	—	—
Meadow fescue :							
S53	17	16	—	—	22·4	18·3	—
S215	—	—	—	32·0	25·0	22·4	15·3
Agrostis	—	—	—	—	20·4	19·3	—

Species and variety of grass

Data from a number of experiments in Britain with various grass species and varieties grown as pure swards are summarized in Table 7. (In 3 of the experiments, the annual rate of N applied varied during their course.) The higher responses to N obtained by Widdowson *et al.* (490) are probably mainly due to less frequent cutting compared with the other experiments. In all but two experiments, responses were highest from cocksfoot. The differences in response between ryegrass, timothy and meadow fescue shown in Table 7 are small. However, on the basis of a large number of experiments, Hunt (243) considered that short-lived swards of Italian and perennial ryegrasses gave better responses to fertilizer N than other grasses. At Hurley, Italian ryegrass showed no greater response than perennial ryegrass over a two-year period (186, 188).

A comparison of 6 grass species grown in the USA has been reported by Cooper *et al.* (105). In general, the most productive species (tall fescue, smooth bromegrass) gave the best responses to fertilizer N, but differences between species were not entirely consistent over a 3-year period.

There are conflicting reports on the importance of differences between varieties within a species. It appears that cultivated varieties bred for maximum production under conditions of plentiful nutrient supply show similar high responses to fertilizer N, but that strains of the same species taken from natural or semi-natural grasslands may show considerably lower responses. A comparison of perennial ryegrass strains illustrating this difference under sand culture conditions has been reported by Antonovics *et al.* (17).

With perennial ryegrass, Lazenby and Rogers (236) found that various clones, although not belonging to specific varieties, varied considerably in efficiency of N utilization and required widely differing levels of fertilizer N in order to achieve similar yields. In the USA, Dotzenko and Henderson (148) found that out of 5 varieties of cocksfoot, one showed a much greater response than the others to rates of N up to 160 lb/acre per year, and that whereas 3 varieties showed almost linear responses up to 320 lb N/acre, the other two showed little response to rates above 160 lb. Cowling and Lockyer (115) found that Irish perennial ryegrass gave a lower response than S23 or S24, in part a consequence of its poor persistence on the high-N plots.

On the other hand, Davies *et al.* (127) found no difference between 3 early perennial ryegrass varieties, S24, Irish and Reveille, in response to N at rates up to 210 lb/acre. Responses of 4 cocksfoot varieties to N rates up to 140 lb/acre also differed little. Weiss and Mukerji (479) also reported no appreciable differences in response between 8 cocksfoot varieties.

Age of sward

Because of the striking improvement in productivity resulting from ploughing and reseeding large areas of poor pasture in Britain during 1939-50, it came to be widely held that recently sown swards almost invariably outyield old permanent pastures. However, subsequent investigations, summarized by Mudd and Meadowcroft (344), indicated that old permanent pastures of good botanical composition could be just as productive as recently sown swards, and that differences in N supply were largely responsible for the yield differences previously recorded. This view has received more recent support from Hunt (243) and Hoogerkamp and Minderhoud (240), who concluded that the decline in productivity with sward age can be overcome by applying N, provided that persistent grass varieties are sown. Hunt reported that the yields of swards given little or no N declined from about the 3rd to about the 7th year after sowing, but that there was no fall in production when 200 lb N/acre per year was applied. The decline in yields of swards receiving low levels of N was attributed entirely to reduced clover growth. Farm-scale trials in the Netherlands have shown no decline in production from permanent pastures given continued application of high levels of fertilizer N (336).

CHAPTER 19

INFLUENCE OF NUTRIENTS OTHER THAN NITROGEN

The yield response of swards to N may be limited by deficiencies in the supply of other nutrients, especially K or P. Although grasses have a greater ability than many other plants to extract K and P from the soil, restrictions of growth caused by deficiencies in these elements are not uncommon in Britain, particularly when high levels of N are used and the herbage is regularly cut and removed rather than grazed. In some parts of the world, deficiencies of P and K are of course widespread and would undoubtedly restrict the response of grass to N, but there appears to be little information from outside Britain on these interactions. No instances have so far been reported in Britain of grassland responses to N being limited by nutrients other than K or P; instances undoubtedly occur in some parts of the world, but quantitative data on such effects appear to be lacking.

Ca deficiency, as such, is not normally a direct cause of poor response to N ; however, acid soil conditions discourage the growth of the more N-responsive grass species, and liming of soils with pH values less than about 5·5 is necessary to achieve maximum efficiency of fertilizer use.

Potassium

In Britain, K deficiency is not uncommon, particularly on light soils, although it should be noted that many grassland experiments have shown no response to applied K, particularly when they lasted only one or two years.

A striking example of reduced N response as a result of K deficiency in mixed grass/clover swards cut for conservation was reported by Holmes and MacLusky (235) and Castle and Holmes (93), working at the Hannah Dairy Research Institute,

Scotland (see Fig. 9). In their trial, 1st-year yields when N was applied alone or in combination with K were similar but, by the 2nd year, soil K supplies where no K had been applied had become so depleted that the yield with applied N was no greater than that without N.

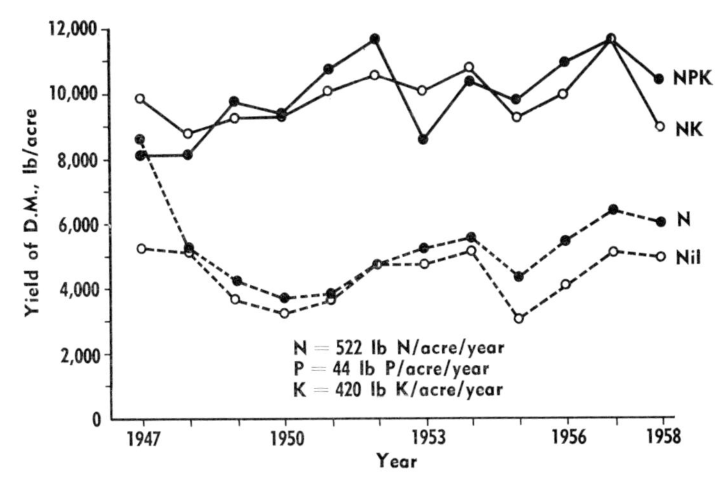

Fig. 9. The influence of K on response to N over a 12-year period (data from Holmes and MacLusky (235) and Castle and Holmes (93)).

Reith *et al.* (395) also reported that K applications considerably increased response to fertilizer N at 5 of the 6 sites investigated in Scotland. The effect of K was greatest at the highest N level of 348 lb N/acre per year and in the 2nd and 3rd years, showing that there was progressive depletion of soil K.

Instances of responses of grass/clover swards to N being increased by fertilizer K have also been reported in England from North Wyke, Devon (513), from the National Institute for Research in Dairying, Reading (350), and from Northern Ireland (98, 298).

In a 6-year experiment at Rothamsted with Italian ryegrass followed by cocksfoot in which the herbage was cut and removed, K had little effect on yields during the first 4 years, but in the 6th year produced a significant effect when N was applied at 270 or 400 lb/acre per year, though not when only 135 lb N was applied (492). Another example of K supply increasingly influencing the response to N over 3 years of an experiment in which the herbage was cut and removed was reported by Hemingway (216). Fertilizer K also increased the response of cocksfoot to N in a field experiment in Indiana, USA (200).

Soils differ widely in their capacity to supply K to plants and therefore in the need to accompany fertilizer N with K. Sward management is also very important. Systems involving the repeated removal of cut herbage also remove large amounts of K from the soil. On the other hand, grazing systems involve the return of much K in animal excreta and depletion is much less rapid.

Phosphorus

In Britain, P deficiency is most marked in the wetter upland areas and swards in such areas usually show large increases in clover growth and total herbage yields following applications of P.

In lowland areas, P deficiency is less widespread than that of K, though, like K deficiency, is most liable to develop where large crops of herbage are cut and removed.

In trials in Warwickshire reported by Gething (177), applications of P greatly increased the response to N of a mixed-grass/clover sward. In the 2nd harvest year the sward was cut 5 times with the results shown in Table 8. The yields quoted are the mean values for the 3 levels of K. Both N and P were applied in 5 equal dressings.

TABLE 8

Herbage yields from grass/clover swards at 4 levels of N with and without P (Gething (177))

	lb N/acre/year			
	0	88	175	350
Yield without P, lb DM/acre	6059	6888	9083	13,227
Yield with 46 lb P, ,, ,,	6529	7840	10,931	17,113

In an experiment in Devon (513) the application of P also markedly increased the response to N of all-grass swards, the effect increasing over a 3-year period.

A large response to P by an all-grass sward was also reported by Wheeler (481) at Wye, Kent. P at 44 lb/acre gave a yield increase of about 30% on plots receiving 275 lb N/acre per year together with sheep excreta. In the absence of P, yields with 275 lb N were no greater than with 160 lb N. There were no zero-N plots in this experiment.

In the experiment at Rothamsted (492), reported above, in which Italian ryegrass, followed by cocksfoot, was sown on an arable soil, a response to P was obtained only at the highest levels of N (100 lb N/acre per cut) and K (56 lb K/acre per cut) and, even then, the average response over 6 years was only 3%. In a further experiment on permanent grass reported by Widdowson *et al.* (494) there was some response to P at all levels of N application, but no consistent effect of P on the response to N.

Similarly, in a 12-year experiment at the North of Scotland College of Agriculture (360), a mixed sward gave an appreciable yield response to phosphate in the last 3 years only and then only at the highest N level used (277 lb/acre per year). It is unlikely that any appreciable quantity of clover would have been present in this sward after 9 years of heavy N applications.

No significant responses to P were obtained in experiments at the Hannah Dairy Research Institute, Scotland, reported by Holmes and MacLusky (235) and Castle and Holmes (93). Also, at 6 sites in Scotland where trials were set up by Reith *et al.* (395), at only two was there any response to P, and these responses were small and only just reached significance.

A harvested crop of 10,000 lb of herbage DM of average mineral composition will contain about 200 lb K and 35 lb P per acre. With both these nutrients, as with any others in short supply, deficiencies are most likely to develop under a system involving the repeated cutting and removal of herbage.

CHAPTER 20

INFLUENCE OF WATER SUPPLY

When fertilizer N is applied to a dry soil, rain or irrigation is necessary to carry it into the root zone; irrigation thus usually increases the response of grass to applied N in dry climates. In humid regions, however, the relationship between N application and irrigation is more variable, and in a number of experiments irrigation has even been reported to decrease response to N. The value of any single irrigation treatment will clearly be influenced by whether or not it is followed within a few days by natural rainfall. The relationship between N supply and irrigation may also be complicated by the presence of clover, which, in the absence of fertilizer N, shows a greater response than grass to irrigation.

A number of experiments in Britain have shown fertilizer N and irrigation to act independently in their effects on annual herbage yields (444), while others have shown positive interactions, i.e. the response to N plus irrigation has exceeded the sum of the responses to each applied independently. Factors which increase the likelihood of a positive interaction between fertilizer N and water supply are: (1) low rainfall on the control plots in irrigation experiments, or during some of the years in year-to-year comparisons; (2) high levels of fertilizer N; (3) absence of clover. The results of several experiments will now be examined with reference to these factors.

Irrigation has been reported to increase the response of all-grass (or predominantly grass) swards to N in a number of investigations in south-east England. Early work by Greenhill (194) at Jealott's Hill, Berkshire, where the average annual rainfall is 26 in., demonstrated that irrigation markedly increased the response to N of average-quality pasture (presumably of mixed composition). The effect of water supply on N response was particularly marked when N rate was increased from 200 to 310 lb/acre per year. With a sward of Italian ryegrass at Reading, D'Aoust and Tayler (123) reported that while irrigation did not increase the response to N over the whole season when the N rate was increased from 112 to 224 lb/acre, it substantially increased it when the N rate was increased from 224 to 448 lb/acre. Both positive and negative interactions occurred over the 6 consecutive 4-week growth periods, and while these were approximately equal over the whole season for the comparison between 112 and 224 lb N, the positive exceeded the negative interactions for the comparison between the 224 and 448 lb N rates. These results taken in conjunction with weather data suggested that the positive effect of irrigation on response to N was due to the improvement it brought about in the moisture status of the top few inches of soil, thus rendering N and other nutrients available to the roots. The importance of an adequate water supply in the topsoil for grass growth has been demonstrated by Garwood and Williams (171).

The report by Penman (378) on work at Woburn in which a cocksfoot sward received different N and irrigation treatments during 1954-9, indicated that only in the dry year of 1959 did irrigation increase response to N. In that year, the two levels of N investigated were approximately 200 and 400 lb/acre. In 1954-6, the rates of N were approximately half these values and so irrigation would have been less likely to increase the response.

At the Hannah Dairy Research Institute, Ayrshire, where the annual rainfall is about 36 in., irrigation did not increase the response to N applied at rates up to 312 lb/acre per year to either grass or grass/clover swards (390).

In experiments in Finland reported by Raininko (385), irrigation increased the response of grass swards to 200 kg N/ha by 2·5–25%, the effect being greatest with cocksfoot. Irrigation did not increase the response to N of grasses in mixture with red clover, except in the case of cocksfoot.

The extent to which rainfall variation and the level of fertilizer N can influence the interaction between N and irrigation is demonstrated by an experiment in Ohio reported by Prine *et al.* (382). Swards of several individual species given various rates of N were either not irrigated, or the soil was brought to field capacity whenever the soil moisture deficit reached 1·5 in. A selection of their results, including calculated DM responses

TABLE 9

DM yields and responses of 4 grass species to levels of N under non-irrigated and irrigated conditions. (Data from Prine *et al.* (382))

		1956				1957			
		Non-irrigated		Irrigated		Non-irrigated		Irrigated	
	N applied, lb/acre/year	Yield, lb DM/acre	Response, lb DM/lb N	Yield, lb DM/acre	Response, lb DM/lb N	Yield, lb DM/acre	Response, lb DM/lb N	Yield, lb DM/acre	Response, lb DM/lb N
Timothy	0	1640	—	2250	—	2830	—	2560	—
	60	2430	13·1	5070	47·0	4830	33·3	4730	36·1
	120	4110	20·6	6580	36·1	5680	23·8	6850	35·8
Cocksfoot	0	1740	—	2470	—	2570	—	3050	—
	60	2940	20·0	4140	27·8	4890	38·7	4640	26·5
	120	4560	23·5	5540	25·6	6190	30·2	6530	29·0
	240	6040	17·1	9400	28·9	9010	26·8	10770	32·2
Bromegrass	0	1760	—	2650	—	2690	—	2710	—
	60	2920	19·3	4480	30·5	4980	38·1	4680	32·8
	120	4200	20·3	6510	32·2	6810	34·3	6360	30·4
	240	5400	15·2	9350	28·0	9970	30·3	9500	28·3
Bluegrass	0	1610	—	1970	—	2520	—	1950	—
	60	1950	5·7	3300	22·1	4670	35·8	3820	31·2
	120	3450	15·3	4890	24·3	6030	29·3	6260	35·9

to fertilizer N for the years 1956 and 1957, is given in Table 9. 1956 was the driest season of the experimental period (1956-8), with rainfall in June, July, August and September below normal. In 1957 rainfall was adequate until the middle of July and, although rainfall in July and August was less than in 1956, the soil at the beginning of July was at field capacity. Rainfall during 1958 was heavy and well distributed. The swards, which were sown in April 1955, were cut 5 times per year at monthly intervals during 1956 and 1957. The data indicated that in the dry year of 1956, irrigation substantially increased N response, but that in 1957 the effects of irrigation were small and inconsistent.

In investigations by Nielsen (354) in which perennial ryegrass was grown in covered lysimeters and received 5 rates of applied water and 5 of fertilizer N, irrigation increased the response to N, particularly at the higher N rates (see Fig. 10). In trials in

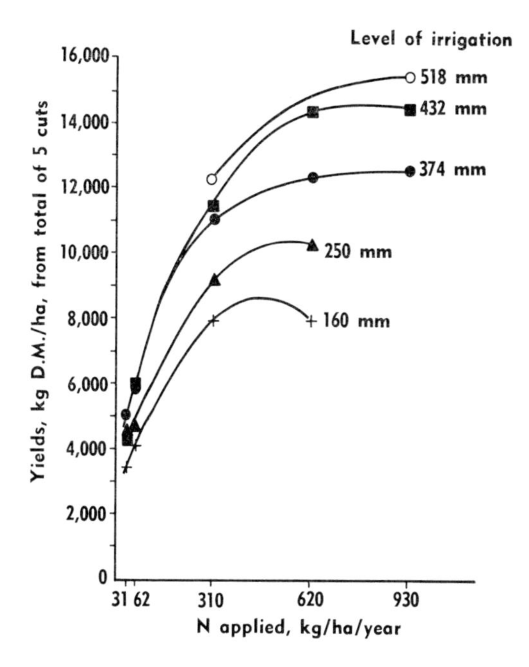

Fig. 10. The response to fertilizer N of ryegrass grown in covered lysimeters at 5 levels of irrigation (from Nielsen (354)).

Alabama, irrigation also increased the response of Coastal Bermudagrass to N at rates up to 400 lb/acre (147).

With grass/clover swards, irrigation normally increases herbage yield independently of N treatment, but may not increase the response to fertilizer N. In fact, the yield increase as a result of irrigation is often greatest on the no-N plots on which clover contents tend to be highest and in such cases there is often a negative interaction between irrigation and fertilizer N.

In experiments carried out over two years at Jealott's Hill, Berkshire, irrigation had no effect on the annual yield response of grass/clover swards to fertilizer N (294). However, irrigation significantly increased the response to N of mixed swards at Reading in both 1955 and 1956, but these swards contained little clover, the soil over at least part of the experimental area was very free-draining and 1955 was very dry (347).

The influence of irrigation on the response of a grass/clover sward to fertilizer N was examined at Hurley by Stiles and Williams (439) during 1956–9. In 1959, the only dry year, there was a positive interaction between irrigation and the highest level of fertilizer N (486 lb/acre per year). Growth on the unirrigated plots was severely restricted by drought, and the fertilizer either remained on the soil surface or, if washed in by light rain, soon became unavailable owing to drying out of the topsoil. In the other years no large moisture deficits developed, and although irrigation always increased herbage yields, it generally reduced the annual response to fertilizer N. Some of the harvests taken at approximately monthly intervals, showed positive and others negative interactions between irrigation and N.

When the water supply is less than that required for maximum yield but is not severely deficient, irrigation and N are to some extent interchangeable. At Woburn, Penman (378) found that with a cocksfoot sward, doubling the N application over the range 17-67 lb/acre per cut had an effect equivalent to adding 0·5 in. of readily available water. With only 17 lb N/cut, growth was checked at a soil moisture deficit of 1 in., whereas with 34 lb N/cut, growth was not checked until the deficit reached 1·5 in. The interchangeability of fertilizer N and irrigation and the extent to which this is influenced by weather conditions is also shown by the data in Table 9. However, it should be noted that, at least in Britain, fertilizer N is much cheaper than irrigation for each unit of extra grass produced (332).

The investigation by Prine *et al.* (382) included a study of the effects of irrigation during the previous summer on the yields of herbage from the first cut in 1957 and 1958. In many instances much of the yield increase during the year irrigation was applied was offset by a fall in yield in the subsequent year, an effect attributed largely to a reduction, caused by irrigation, in the supply of residual N available in the following spring.

It is clear that for maximum herbage production, swards receiving high inputs of fertilizer require more water than do those receiving low inputs. Irrigation often enables grass to respond to a higher rate of fertilizer N. In some situations it may, however, result in an increased loss of N by leaching. With moderate applications of fertilizer N, irrigation may increase herbage yield but have little effect on the actual response in terms of lb DM produced per lb of N applied.

A shortage of water in the topsoil appears to restrict grass growth by reducing the availability of nutrients to plant roots even when sufficient water for maximum growth is available to the roots at lower depths.

CHAPTER 21

INFLUENCE OF HIGH RATES OF NITROGEN APPLICATION AND OF SOIL NITROGEN STATUS

Many experiments in temperate regions have shown an almost linear response to applications up to 300–400 lb N/acre per year for both all-grass and grass/clover swards (see pp. 69-70).

Experiments in which higher levels of N have been applied have generally shown a decline in response to rates between 300 and 600 lb N, and a decrease in yield at some point between 600 and 1200 lb N/acre per year (18, 156, 190, 242, 249, 406). A selection of data from these experiments is shown in Fig. 11. The level of N supply to which there is a response in terms of annual yield is higher when defoliation is relatively frequent (230, 242). Also, the more favourable the conditions for grass growth, the higher the level of N to which there is a yield response. Since the response to fertilizer N is usually linear up to an annual input of about 350 lb N/acre per year, variations in soil N status are unlikely to affect response per lb of applied N unless the contributions from soil plus fertilizer amount to somewhat more than this quantity. The actual yields obtained, particularly on plots not receiving fertilizer N, will of course be influenced by soil N status. As indicated in Chapter 8, the quantity of N harvested in the herbage from all-grass plots not receiving fertilizer N may range from about 10 lb/acre per year for some soils previously under arable cultivation and low in OM, to more than 100 lb for some soils under permanent pasture.

No experiments appear to have been carried out to examine specifically the effect of soil N status on response to fertilizer N. A number of workers (63, 114, 473) have, however, investigated the statistical relationship between the yield of herbage DM and/or N yield and the estimated total N supply contributed by fertilizer N, soil N and, in appropriate circumstances, N fixed by

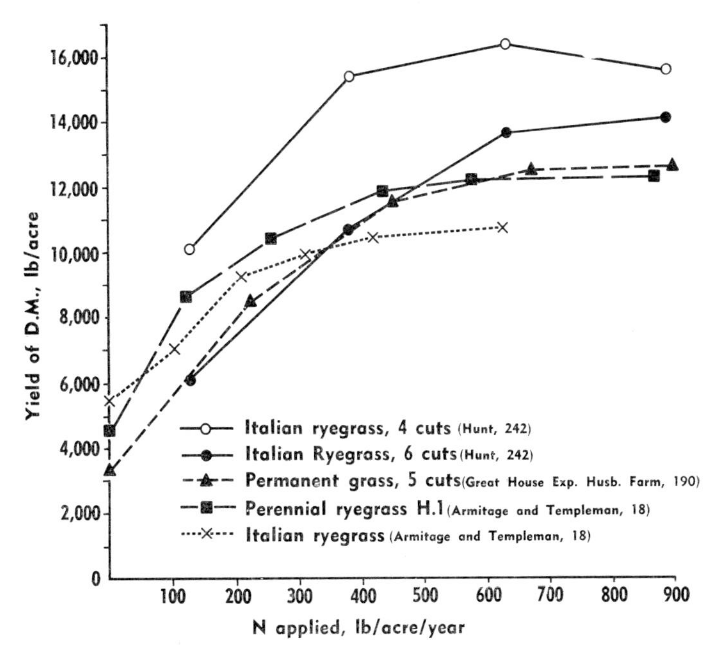

Fig. 11. Responses of grass swards to very high levels of fertilizer N.

clover and that returned in animal excreta. Apart from fertilizer N, these items are all difficult to assess accurately. Means of estimating the quantities of N fixed by clover and returned in animal excreta are discussed in Chapters 2 and 5, respectively. For soil N, there is no reliable chemical method for assessing the supply available to an established grass sward (see Chapter 26). From their examination of published data, Walker *et al.* (473) concluded that there was a good correlation between the total N content of the soil and the yield of N in various crops, and that 0·1% N in a soil would yield about 25 lb of mineral N/acre per year, of which about 17 lb would be harvested. However, in experiments with all-grass swards at 24 sites in Britain, Brockman (63) found that although soils with a low total N content (0·1–0·2%) tended to give lower herbage N yields than soils with a higher content (0·4–0·5%), the relationship was not sufficiently close for total soil N to be of much value in estimating available soil N. He considered that the herbage N yield from plots not receiving fertilizer N was a better value.

Although soil N status can influence herbage yield, at least at rates of fertilizer N below about 300 lb N/ acre per year, its effect on response to fertilizer N appears to be relatively small in comparison with such factors as weather and defoliation frequency.

CHAPTER 22

INFLUENCE OF THE CHEMICAL FORM OF FERTILIZER NITROGEN

The most commonly used forms of fertilizer N are ammonium nitrate (alone, or combined with calcium carbonate in fertilizers such as Nitro-chalk), ammonium sulphate, ammonium phosphate (in many compound fertilizers), calcium nitrate, ammonia (anhydrous or in aqueous solution), and urea (in some compound fertilizers). The use of fertilizer solutions and slow-release nitrogenous fertilizers for grass has also been investigated, but these forms have major shortcomings. Of the organic manures, slurry and gülle (slurry diluted with water) are the most important types for grassland, although farmyard manure is also used in predominantly pasture areas.

Uptake of fertilizer N by plants in the field occurs mainly in the form of nitrate, but where urea, ammonia or its salts are applied, an appreciable proportion may be absorbed as ammonium ions. That the ammonium form of N is readily absorbed by grass is indicated by the finding of Richardson (399) that the disappearance of ammonium from the soil, following the application of ammonium sulphate, was almost as rapid as the disappearance of nitrate after the addition of sodium nitrate. Plants can also absorb urea. However, differences in yield response to the various forms of fertilizer N appear to be due mainly to differences in the extent of N loss during the interval between application and uptake, rather than to differences caused by the form in which the N is actually absorbed. The method of application and the occurrence of toxic effects may also be important.

Ammonium and nitrate salts

Ammonium and nitrate salts generally produce yield responses as high as, or higher than, urea, ammonia and organic forms of N. The relative merits of ammonium and nitrate depend largely on the soil and climate which, together, govern the extent of gaseous and leaching losses. Gaseous losses may take the form either of ammonia (from ammonium salts), or of the products of denitrification (from nitrates and from ammonium salts after nitrification). Loss of ammonia from ammonium salts is particularly important when they are applied to calcareous soils in warm conditions. In addition to the effects of soil pH and temperature, losses may also be influenced by the associated anion. Larsen and Gunary (283) reported a greater loss from ammonium sulphate than from mono- and di-ammonium phosphate and ammonium nitrate. These factors probably explain the results of Devine and Holmes (134), who compared single applications of ammonium sulphate and ammonium nitrate at rates of 30–100 lb N/acre in 89 experiments on grassland. On most soils, the two forms produced similar responses, but on soils containing more than 10% calcium carbonate, and at the 100-lb N per acre rate, ammonium sulphate gave only 79% of the response given by ammonium nitrate. Heddle (213) reported that Nitro-chalk and ammonium sulphate were equally effective for Italian ryegrass on various sites in eastern Scotland. Nitrate N is more susceptible than ammonium N to gaseous losses of N_2 and nitrogen oxides, but such losses can occur from ammonium salts after nitrification (see Chapter 10). In experiments with ^{15}N, Woldendorp (506) found that 15–37% of the nitrate N added to cores of permanent grassland was lost in gaseous form after two months. Losses from ammonium N were rather less than half those from nitrate N.

Losses of N through leaching are rarely significant in grassland during the growing season. However, where conditions favour leaching, considerable amounts of nitrate may be lost, whereas ammonium, except on sandy soils with a low cation-exchange capacity, is not susceptible to leaching until nitrified (232). Devine and Holmes (135) reported that when applied in early winter and particularly when followed by a high winter rainfall, calcium nitrate gave lower yields of grass than ammonium sulphate, ammonium nitrate and urea. However, all these forms of fertilizer N gave lower yields when applied in early winter than when applied in late winter or spring, the difference being greater when winter rainfall was high. Mc-

Allister *et al.* (306) noted that Nitro-chalk produced 12% less response than ammonium sulphate when both were applied on about 22 February, but similar responses when applied on about 21 March. Leaching was probably involved in both these investigations, but denitrification may also have contributed to the lower responses obtained from nitrates.

Most comparisons of different forms of fertilizer N have been made in trials lasting only one or two years. However, the secondary effects of a given form of fertilizer N may become important for long-term responses.

It is well known that the application of ammonium salts and of compounds yielding ammonium salts increases soil acidity and hence increases also the need for lime. This is due to nitrification of the ammonium ion. The application of ammonium sulphate to soil will result in the following sequence of reactions :

Soil colloid
Ca
Ca
$Mg + 2NH_4^+ + SO_4^{--}$ $\xrightarrow[\text{exchange}]{\text{cation}}$ NH_4
Ca | Ca
Ca | Mg
| $NH_4 + Ca^{++} + SO_4^{--}$
| Ca
| Ca

NH_4
Ca
Mg $\xrightarrow{\text{nitrification}}$ H, Ca, $Mg + 2H^+ + 2NO_3^-$, H, Ca, Ca
NH_4
Ca
Ca

H
Ca
$Mg + 2H^+$ $\xrightarrow[\text{exchange}]{\text{cation}}$ H, Ca, $Mg + Ca^{++}$, H, Ca, H, H
H
Ca
Ca

Although two Ca ions are shown as being replaced on the soil colloid by 4 H ions, Mg, K, Na and trace element cations are all liable to be displaced and are then subject to leaching. If there were no loss of gaseous ammonia or uptake of ammonium N, then the application of 100 lb N as ammonium sulphate would result in the displacement of cations equivalent to 700 lb calcium carbonate. Cooke (104) has estimated that, in practice, there is a loss of the equivalent of about 500 lb calcium carbonate per 100 lb N applied as ammonium sulphate. The loss due to cation exchange and nitrification is probably less than the equivalent of 500 lb calcium carbonate on grassland owing to the substantial uptake of ammonium (see p. 89). However, when ammonium is absorbed by plants there is a physiological acidifying effect on the soil due to the ammonium being absorbed in excess of the accompanying anions, the excess being, in effect, balanced by an increase in H ion concentration. Conversely, when nitrate is absorbed there is an increase in soil alkalinity equivalent to the excess absorption of nitrate over the associated cations.

Assessments of the relative merits of the various forms of fertilizer N should include consideration of the need to restore the cations displaced by the reactions of ammonium described above. The restoration of Ca by liming is often, on non-calcareous soils, the main requirement. With Nitro-chalk (21% N) about half the acidifying effect of the ammonium is counteracted by the calcium carbonate present.

The respective effects on grass growth of N taken up either as nitrate N or ammonium N were investigated by Nielsen and Cunningham (355) in glasshouse experiments in which Italian ryegrass was grown in soils containing 6 levels of N (0–500 ppm.) and with the nitrification inhibitor 2-chloro-6-(trichloromethyl) pyridine ‘ N-Serve ’* used to prevent bacterial nitrification of ammonium N. With nitrate N, herbage yields were highest at the 100-ppm. level, and with ammonium N at 200 ppm. At 100 ppm., yields with nitrate were about 14% higher than with ammonium, but at 200 ppm. and above, ammonium resulted in better yields than nitrate, the difference increasing from about 2% of the nitrate N value at the 200-ppm. level, to about 30% at the 500-ppm. level.

* (Registered Trade Mark, Dow Chemical Co.)

Urea

The advantages and disadvantages of urea as a fertilizer have been reviewed by Gasser (174).

In soil, urea is hydrolysed to ammonium carbonate. There is an attendant rise in soil pH which, if it exceeds a value of 7, leads to the formation of free ammonia which may then be lost to the atmosphere. Hydrolysis of urea is most rapid in soils with high OM content and is thus likely to proceed more rapidly in soils under grass than in arable soils (174).

Low and Piper (296) have shown that under cold, moist conditions typical of those of an English spring, 90% of an application of urea would be converted to ammonium in two days and of this only about 14% would be nitrified.

Gaseous losses of 20–30% of the N in urea applied to established grass swards on somewhat acid sandy soils were reported to occur within 7 days by Volk (462). Templeman (445) reported that for equivalent rates of N, the yield response of grassland to urea at 24 sites in England averaged about 15% less than that to Nitro-chalk. Although at most sites yield responses to the two forms of N did not differ significantly, urea gave much lower yields at a small number of sites On the basis of numerous field trials in the Netherlands, Dilz and van Burg (138) reported that urea applied to grassland at 60 kg N/ha was, on average, only 80% as effective as nitro-lime ; with urea, 10–17% of the N was usually lost within 3 days, although the loss was reduced if rain fell soon after the application. In pot experiments, Low and Piper (296) found that when top-dressed on established grass, urea was slightly less effective than Nitro-chalk or ammonium sulphate, but all 3 fertilizers were equally effective when mixed with the soil before sowing. This difference is probably explained by a loss of ammonia from the urea top-dressing.

In a number of investigations in the USA, losses of N as ammonia from applications of urea to grassland have ranged from 2 to 36% (174).

Urea leaches almost as readily as nitrate (60) but, as stated above, is normally converted rapidly to ammonium carbonate.

In some of the earlier fertilizer trials, the urea used contained phytotoxic amounts of biuret. According to Cooke (104), 1%

of biuret in urea is insufficient to injure plants, and this level is unlikely to be exceeded in the urea currently manufactured. Templeman (445) found no differences in the response of grass swards to urea with biuret contents in the range 0·13–2·65%.

The ammonia released on hydrolysis of urea appears to present a hazard to germinating seedlings (296). This probably explains why urea has been observed to cause a greater check to establishment of newly sown Italian ryegrass than ammonium nitrate, ammonium sulphate or calcium nitrate, as reported by Widdowson *et al.* (492).

Although urea causes an initial rise in soil pH, it subsequently tends to acidify the soil as a result of nitrification.

Anhydrous and aqueous ammonia

Following the development of improved injection equipment, increased attention has been given to the use of anhydrous and aqueous ammonia on grassland.

Since ammonia is the primary product of the synthetic N fixation process, it is considerably cheaper, per unit of N, than the various nitrate and ammonium salts and urea. With 82% N, anhydrous ammonia is the most concentrated form of fertilizer N possible. It is available in liquefied form under pressure. The commercially available solutions of aqueous ammonia usually contain 29% N, sometimes less, and are not under pressure.

Because it is necessary to inject both anhydrous and aqueous ammonia about 4 in. below the soil surface in order to minimize gaseous losses (78), application costs and, on some soils, the inefficiency of the injection equipment partly offset their low initial cost.

It has been suggested that a single application in early spring might provide a steady supply of available N throughout the season, thus eliminating the need for successive applications for each defoliation. This concept is based on the view that the ammonium ions resulting from the reaction with the soil water will be adsorbed by the soil in a zone 2-6 in. on each side of the injection line (176), that the high concentration of ammonia will kill roots and inhibit nitrifying bacteria along the injection line, and that uptake will therefore proceed steadily from the periphery of the adsorption zone. A further suggested advantage of ammonia is that it can be applied during winter or early spring because adsorption virtually eliminates leaching and denitrification losses.

In some of the field trials reported, ammonia applied in one large dose in winter or early spring has been compared with solid N fertilizers applied in smaller amounts at intervals during the growing season. In other trials, ammonia and solid fertilizers have been applied at equal rates on the same dates.

In many of the comparisons made between single applications of ammonia and split applications of solid fertilizer, yields have been 10–40% lower with the ammonia (77, 78, 79, 255, 256, 487). However, Hodgson and Draycott (227) found that aqueous ammonia injected in early April was as effective as split applications of solid fertilizer if more than 400 lb N per acre was applied.

An experiment on permanent grass at Rothamsted in 1967 also showed that, with a rate of 448 lb N/acre, a single application of aqueous ammonia in either spring or autumn was as effective as the split application of the same total rate of Nitro-chalk. Anhydrous ammonia was rather less effective than aqueous ammonia, particularly at high rates (488).

In comparisons between ammonia and solid fertilizers applied at equal rates of N on the same dates, the results are influenced by the number of cuts taken per application. Ammonia and solid fertilizer both supplied as large single applications in spring have often given similar total annual yields (2, 78, 79, 164, 488). The ammonia produced lower yields at the first cut, but rather higher yields at subsequent cuts. The relatively slower uptake of ammonia was confirmed in the experiment reported by Jeater (255) in which ammonia and ammonium nitrate were each applied at two rates (67 and 135 lb N/acre), two cuts of herbage taken, and the same N applications and harvesting procedure repeated. Ammonia produced lower yields than ammonium nitrate at the 1st and higher yields at the 2nd cut. Similar results were obtained by Crooks *et al.* (117) using rates of 40 and 80 lb N/acre. Ammonia gave lower yields than solid fertilizers when both were applied at 92 lb N/acre in 4 experiments reported by Jameson (252) in which two cuts were taken after each application.

In an experiment in which ammonia and nitro-lime were applied for each cut and in which the sward receiving nitro-lime was damaged by the passage of injection tines to the same extent as that receiving ammonia, van Burg *et al.* (78) found considerably higher yields from the nitro-lime. This confirmed the earlier results of van Burg and van Brakel (77).

The results of the field trials summarized above indicate that ammonia has often given lower yields than conventional solid fertilizers, except in the unrealistic type of comparison in which only one application of solid fertilizer containing a soluble compound of N is made during the season.

Possible reasons why ammonia has been less effective than solid fertilizers in some investigations are :

(1) damage to the sward caused by injection equipment,
(2) loss of gaseous ammonia under certain conditions,
(3) release of available N from adsorbed ammonia being less well matched to the requirements of the sward than regular top-dressings of solid fertilizer.

A single passage of injection tines in the absence of any fertilizer was found to cause a small reduction in yield in one experiment by Jeater (255), but not in a later experiment by the same author (256). Van Burg and van Brakel (77) found that damage from tines increased as the depth of injection increased from 5 to 15 cm, and was considerably greater on a clay soil than on sandy or peat soils. Jameson (252) also found that damage was greater under dry than under moist soil conditions.

The extent to which ammonia is lost depends on the depth of injection, the efficiency of the equipment, soil type and soil moisture status. On sand and peat soils, van Burg and van Brakel (77) found that ammonia gave slightly higher yields when injected at a depth of 10 cm than at 5 cm and a higher yield at 5 than at 15 cm, but that on a clay soil the yield obtained with injection at a depth of 5 cm was 13% higher than that at 10 cm. The lower yields at shallow injection depths on the sand and peat soils were apparently due to loss of gaseous ammonia. The desirable features of injection equipment for grassland have been enumerated by van Burg *et al.* (78). The efficiency of injection is considerably reduced by the presence of large quantities of stones in the soil (117).

Evidence that the release of available N from adsorbed ammonia is not well matched to the requirements of the sward was obtained by Cowling (112) from an experiment in which aqueous ammonia was injected at fixed points in the soil using a hypodermic syringe, thus avoiding sward damage and the loss of gaseous ammonia. Even with this technique, aqueous ammonia injected in late winter was no more effective than a single application of ammonium nitrate in providing a continuing supply of N throughout the season, and considerably less effective than split applications of ammonium nitrate.

The results obtained by Jeater (256) indicated that when ammonia was injected as early as mid-November, responses were considerably influenced by winter rainfall. At Jealott's Hill, with a rainfall of 13·2 in. between the November injection and the first cut, the yield response to ammonia (300 lb N/acre) was only 64% of that to split applications of ammonium nitrate, whereas at Henley Manor, Somerset, with a rainfall of 23·3 in. during the same period, the yield response to ammonia was only 42% of that to the same quantity of N applied in split applications of ammonium nitrate. Nitrification of the ammonia, followed by leaching of nitrate, probably accounts for the difference between the two sites.

Ammonia is not suitable for top-dressing grassland, as even solutions diluted to 5% N cause damage to the herbage (2, 361).

In addition to frequently producing a lower yield response than solid fertilizers, ammonia also produces herbage with lower contents of Ca and Mg than does Nitro-chalk (77). In common with ammonium salts and urea, ammonia increases soil acidity after nitrification has occurred.

Solutions of ammonium and nitrate salts and urea

A few workers have compared forms of N applied in solution with the same compounds applied as solids. In no instance did the dissolved form produce a higher yield than the corresponding solid, except where the solution was injected into the soil.

In two field experiments in which ammonium nitrate, ammonium sulphate, calcium nitrate and urea were all applied both as top-dressings in solid form and as 5% N solutions sprayed on to the sward, the method of application had no appreciable effect on yield. However, the grass absorbed more N, and herbage N contents were higher, with the solid forms than with the solutions (361). Jameson (252) found that a solution containing ammonium nitrate and urea sprayed on to the sward surface gave somewhat lower yields than top-dressing with solid ammonium sulphate and Nitro-chalk. When injected into the soil, liquid fertilizers also generally gave lower yields than solid top-dressings, partly as a result of mechanical damage caused by the injection equipment (252). Van Burg (74) reported that solutions of ammonium nitrate and urea and a mixture of the two sprayed on to mixed grass swards were all considerably less effective than top-dressings of solid nitro-lime.

Preliminary work reported by Draycott *et al.* (150) indicated that solutions of ammonium sulphate and ammonium nitrate + urea, when injected 3–4 in. deep at more than 200 lb N/acre, may give higher total annual yields than the same amount of N supplied as split applications of solid fertilizer. The advantage of injected solution was greater during dry periods.

Slow-release nitrogen fertilizers

The problem of providing a uniform supply of N over a period of months from a single application is not satisfactorily overcome by the injection of ammonia. An alternative approach is to use a solid fertilizer which releases available N slowly.

Means by which this can be achieved include (1) coating soluble forms of N such as ammonium nitrate or urea with membranes which are semi-permeable, perforated, or able to be degraded by soil micro-organisms, and (2) using more complex N compounds which either have low solubility or are insoluble until broken down by soil micro-organisms (375).

A range of materials, including coated urea and several organic N compounds, was examined by Beaton *et al.* (36) in a one-year growth-chamber experiment with cocksfoot. Yield and N uptake in the 1st harvest were greatest with ammonium nitrate, urea, urea plus thiourea, and finely divided oxamide. Glycoluril ($C_4H_6N_4O_2$) and coated urea gave the highest yields and N uptakes in the 2nd and 3rd harvests. At later cuts, yields and N uptake from the urea-formaldehyde and thiourea treatments increased and, in the last 4 harvests, the yields obtained with these two N sources were among the highest.

Field trials with Kentucky bluegrass showed that, with a single application of 140 lb N/acre, a mixture containing about half the N as urea-formaldehyde and half as ammonium nitrate produced annual yields as high as ammonium nitrate alone and gave more uniform growth through the season (270). Urea-formaldehyde had an appreciable residual effect for at least two years after application.

Urea-formaldehyde alone produced, over one year, little more than half the yield response produced by Nitro-chalk in Italian ryegrass when both forms of N were given as single applications in spring and compared even less favourably with Nitro-chalk applied in split dressings. Maximum recovery of N during two years cropping with grass and one year with barley was 54% for urea-formaldehyde and 90% for equivalent Nitro-chalk (489).

Formalized casein (a waste product of the plastics industry) has also been investigated as a source of N for Italian ryegrass (495). It released little available N early in the season of application but considerable amounts later in the season and some in the following year. Aggregate yields over one or two years were, however, lower than from split applications of ammonium sulphate, calcium nitrate or urea.

The high cost of urea-formaldehyde and other slow-release N fertilizers is the main reason why they are not used on grassland.

Organic manures

With the trend towards increased housing of livestock and stricter legislation regarding the disposal of farm effluent, considerable quantities of slurry and gülle are now applied to grassland.

Grass usually responds less to the N in slurry or gülle than to fertilizer N. However, if only the soluble N (derived from urine) is considered and if the application is made under favourable conditions, the response can be similar to that from fertilizer.

In trials in Scotland by Herriott *et al.* (222, 224, 225), responses from gülle applied to mixed swards were 55–84% of those from equivalent amounts of N applied as Nitro-chalk. The degree of response appeared to be positively correlated with the proportion of urine N. In the most recent trials by these workers, gülle produced a poor response at one site on a sandy soil subject to leaching. McAllister (302) found that the average response of Italian ryegrass to slurry was 66% of that to ammonium sulphate. In a subsequent experiment (303), the responses to N in slurry were 40-60% of those to the N in ammonium sulphate, the highest value being attained with single applications of 240 lb N/acre in spring. When calculated on the basis of soluble-N content, the responses to slurry were very similar to those from fertilizers.

The effectiveness of the N in diluted urine was demonstrated by Castle and Drysdale (92), who obtained responses from a grass/clover sward similar to those given by equivalent rates of N (75 and 150 lb/acre) as Nitro-chalk. The plots receiving the liquid manure had higher clover contents, a factor which may have contributed to the result. Subsequent work at the Hannah Dairy Research Institute (151) showed that liquid manure applied during winter, even as early as November, could give substantial yield responses on a variety of swards and was only slightly less effective than conventional fertilizers. Later, Drysdale (152) showed that liquid manure applied in 5 equal dressings supplying up to 400 lb N/acre per year to grass/clover swards resulted in much greater proportions of clover in the herbage, but herbage yields were almost identical to those given by similar quantities of N as Nitro-chalk.

The application of liquid manure or slurry during winter is likely to result in some loss of N by leaching, and on poorly drained land the sward is liable to be damaged by tractor-drawn distribution equipment. Liquid manure and slurry may also lose N by the volatilization of ammonia during storage.

CHAPTER 23

INFLUENCE OF THE RETURN OF ANIMAL EXCRETA

Under grazing conditions, 75–95% of the N consumed by livestock may be returned to the sward in excreta (see p. 21). As this return enables some of the fertilizer N to be utilized more than once during a growing season, it might be expected that responses to fertilizer N would be considerably greater under a grazing than under a cutting regime. However, recirculation of N by grazing animals involves large losses (see Chapter 6) and, in practice, the influence on herbage production of N returned in excreta is often small.

Interpretation of investigations on the influence of the return of animal excreta is frequently complicated by additional factors. For example, in some comparisons of cut and grazed swards, the frequency of defoliation has differed and is likely to have affected yields appreciably (see Chapter 24). Where K is deficient, any positive effects of return of excreta on response to fertilizer N are likely to be due to the recirculation of K rather than N. Where clover is present, the return of N in excreta may depress clover growth and hence N fixation, thus giving a smaller response to excreta than would occur with an all-grass sward. With grazed swards, treading and selective grazing and fouling by excreta may introduce further complications. The magnitude of these effects of grazing is very much influenced by soil characteristics, weather and grazing management. It has been observed that, as the biological activity of the soil increases as a result of repeated grazing at high stocking rates, sward damage caused by fouling is reduced (234). Differences in the extent of these deleterious effects may provide some explanation for the differences reported in the net effect of grazing on herbage yields.

Grass swards

In experiments carried out by Armitage and Templeman (18), grazed swards yielded an average of about 650 lb DM/acre per year more than cut swards where fertilizer N was not applied, but showed a slightly lower response than cut swards to the application of 350 lb N/acre per year. However, in this investigation the frequency of defoliation was not the same for the two swards.

In a similar experiment reported by Brockman and Wolton (64), grazed swards yielded 1000–1600 lb DM/acre per year more than cut swards and showed a slightly greater response to fertilizer N at rates up to 120 lb/acre. The plots in this investigation were defoliated when the herbage yield, visually assessed, reached about 1100 lb DM/acre. Later work in which the plots were defoliated at the same times showed no difference in yield between cut and grazed plots receiving no N, but the response to rates up to 180 lb fertilizer N/acre was greater on the grazed plots (420). The yields from the no-N plots were however relatively high (4400–6000 lb DM/acre) in this investigation.

At Hurley, the effects of grazing returns on N uptake and herbage yields have been small, even on grass-dominant swards. Green and Cowling (193) reported that even with the application of 280 lb N/acre per year, grazing produced only a slight increase in the uptake of N by herbage.

Grass/clover swards

In studies on the effect of grazing returns on grass/clover swards at Wye, Kent, Watkin (475) and Wheeler (480) found no effect on yield in the absence of fertilizer N, but a marked increase when fertilizer N was applied and clover was therefore of little significance in supplying N. There was, however, evidence of K deficiency on the plots not receiving excreta and the effect at the higher rates of fertilizer N may have been partly due to the return of K.

In experiments reported by Armitage and Templeman (18), grazed swards gave higher yields than cut swards, despite more frequent defoliation on the grazed swards, but there was little difference between the two types of management in the response to fertilizer N at rates up to about 450 lb N/acre per year.

A similar result was obtained by Brockman and Wolton (64), who applied up to 120 lb N/acre per year on their grazed plots. In a later experiment, grass/clover swards receiving up to 180 lb N/acre per year were either cut or grazed with the same frequency of defoliation (420). Yields of herbage DM and N were greater from the cut than the grazed swards where no fertilizer N was applied, but responses to fertilizer N were greater from the grazed swards.

CHAPTER 24

INFLUENCE OF DEFOLIATION INTENSITY

Intensity of defoliation is a function of frequency and height of cut. Work in Britain has indicated that maximum DM yields are obtained with a combination of long periods of uninterrupted regrowth and low cutting height (317). However, more frequent defoliation is often desirable in order to improve herbage digestibility and, in climates which provide greater inputs of light energy, a greater height of cut may be more advantageous. Also, there are likely to be differences between species in the optimum frequency and height of defoliation (317).

Relatively few experiments have investigated the influence of intensity of defoliation on the response to fertilizer N and most of these have included variations only in frequency of defoliation.

Both ryegrass and ryegrass/clover swards gave higher yields with 3 cuts than with 6 cuts per year, and with 6 cuts per year than with 2 cuts per week (184). Height of cut was about 2 in. with the 3 and 6 cuts per year treatments, and 1 in. with the 2 cuts per week treatment. The results obtained for the second year of the experiment with ryegrass alone are shown in Fig. 12. As well as giving the highest yields at all levels of N, the least frequent cutting treatment showed the greatest DM response to N at levels up to about 250 lb/acre per year, but a lower response to the second 250-lb N increment.

Data given by Wolton (512) comparing two defoliation frequencies are shown in Fig. 13a. Although yields were higher with less frequent defoliation, there was no effect on response to fertilizer N at rates up to 200 lb N/acre.

The effect of cutting frequency on response to fertilizer N has also been investigated by Holliday and Wilman (230). With 10 cuts per year in 1958, the yield of grass/clover herbage increased linearly with increasing N up to 417 lb N/acre; with

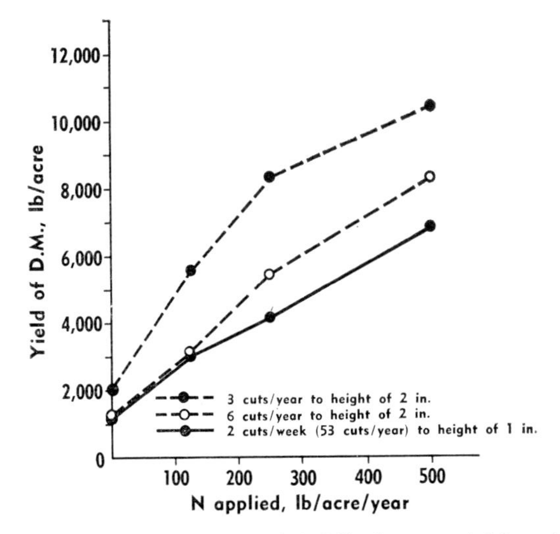

Fig. 12. Influence of frequency of defoliation on yield and response to fertilizer N of S24 perennial ryegrass (184).

6 cuts per year, response to N was still positive but tended to fall off at 417 lb, and with only 2 cuts per year actually declined at N rates above 139 lb (see Fig. 13b). However, at all levels of N, yields with 2 cuts were greater than with 6, which in turn were greater than with 10.

In experiments during 10 years in W. Scotland (242), in which Italian ryegrass was given 127, 380, 633 and 886 lb N/acre and cut 4 times at 6-week intervals each season, the average DM yield was highest (16,400 lb/acre) for the 633-lb treatment. Where 6 cuts were taken at 4-week intervals, yields were lower than where 4 cuts were taken, but continued to increase as N rates increased up to 886 lb/acre. The response curves from this experiment are shown in Fig. 11 (p. 87).

Armitage and Templeman (18) reported that yields from grass/clover swards receiving up to 800 lb N/acre per year and cut to a height of 1·5 in. were 4–7% greater when cut at the silage stage than when cut more frequently at the grazing stage.

The complications introduced into the interpretation of grassland experiments by differences in the height and frequency of cutting have been discussed by MacLusky and

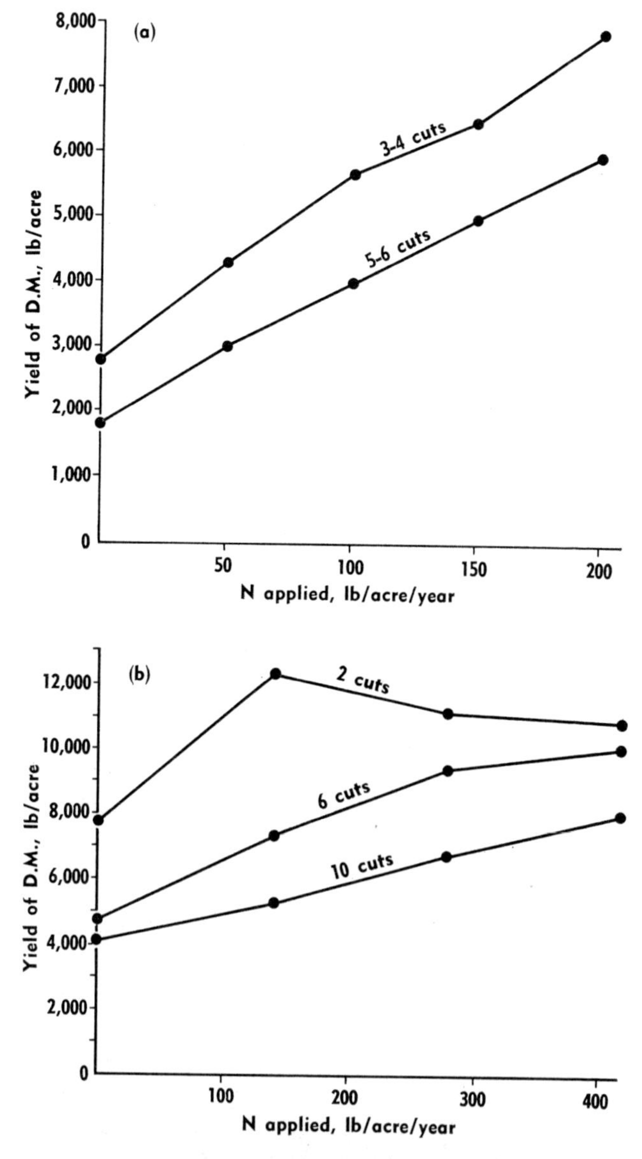

Fig. 13. (a) The influence of frequency of defoliation on yield and response to fertilizer N of an all-grass sward (512).

(b) The influence of frequency of defoliation on yield and response to fertilizer N of a grass/clover sward (230).

Morris (317) and by Wilson (500). As mentioned above, few data are available on the effect of cutting height on response to fertilizer N.

Wilson (500), working in Canada, has reported an experiment in which cocksfoot produced higher yields with a 2-in. than with a 4-in. cutting height at each of 3 N levels, but the yield response to 200 lb N was the same at both cutting heights.

The data summarized in this chapter support the view that under conditions similar to those in Britain, maximum yields and responses to fertilizer N are likely to be obtained with infrequent defoliation and a low height of cut. The optimum frequency and height in practical situations will, of course, be influenced by the quantity and digestibility of herbage required and by the level of fertilizer N to be applied.

CHAPTER 25

INFLUENCE OF SEASON OF THE YEAR AND LENGTH OF GROWTH PERIOD

The response of grassland to fertilizer N is influenced by the time of application, the greatest response, particularly to high levels of N, occurring when seasonal grass growth is at its most vigorous. However, the growing season can be extended by the use of N and, in cool temperate regions, the value of grass produced in early spring or late autumn may justify the use of N at those times, even though the yield response is relatively low. On the other hand, since the response is very dependent on weather conditions, the current use of fertilizer N may be a less efficient means of obtaining herbage for use in early spring and late autumn than the conservation of grass grown in mid-season (111).

In areas where the climate is warmer than in N.W. Europe, fertilizer N may stimulate the growth of grass in winter. Thus in S. Victoria and some other parts of Australia, winter temperatures are generally too low for clover growth and for mineralization of appreciable amounts of soil N, but grass will grow if supplied with fertilizer N (101, 352).

During periods of rapid growth, most of the uptake of applied nitrate or ammonium N takes place during the 3-4 weeks after its application. Only when growth is restricted by cold or drought is uptake likely to be delayed. However, DM production proceeds more slowly than N uptake, so that the yield response to N for about 4 weeks after the application is relatively low.

Season of the year : experiments in Britain and the Netherlands

Because of variations in seasonal weather pattern from year to year, it is impossible to make accurate generalizations regarding seasonal differences in response to fertilizer N. The effects of light intensity and drought have been described in Chapters 16 and 20. Even within a given year, the assessment of differences in seasonal response is difficult. When the same plots are used throughout the year, the residual effect of previous applications of N may be appreciable.

The residual effect may be either positive, resulting from incomplete utilization of N applied for the previous cut, or negative, resulting from the suppression of developing tillers by a very heavy previous crop. However, attempts to eliminate residual effects by not applying N until the growth period to be tested is about to begin, create an artificial situation in which the morphology of the sward differs from that in plots which have received N for each harvest prior to the test period. In the experiments reported, some workers have used the same plots for successive N applications throughout the year and others have used fresh plots for each application. The results of experiments comparing seasonal responses are also influenced by the frequency of defoliation employed, as shown by the work of van Burg (73) discussed below.

The effects of applying N at rates up to 375 lb/acre per cut to permanent grass swards on clay and peaty soils, using the same plots for each of 5 or 6 cuts, were studied by Mulder (345). The results obtained on the two soils differed considerably, but both showed a decline in yield and N response after mid-June. The reduction was much greater on the clay soil where the swards suffered severely from lack of water, but there was a marked recovery on this soil in the period 29 July–19 August, when yields for levels up to 160 lb N exceeded those obtained in the first two periods up to mid-June. Although the reductions in yield and N response were less on the peaty soil where water supply was better, there was no recovery during August. Mulder attributed the seasonal differences partly to the weather and partly to the physiological condition of the plants.

Cowling (108) studied the response of cocksfoot to N applied 4 times each year during the period 1956-9. In order to reduce residual effects, the N was applied either at the beginning of the 1st and 3rd growth periods or at the beginning of the 2nd and 4th periods. The results are shown in Fig. 14. Data for 1956-8 are averaged, since the pattern of growth in these years was similar. Results for 1959, an extremely dry year,

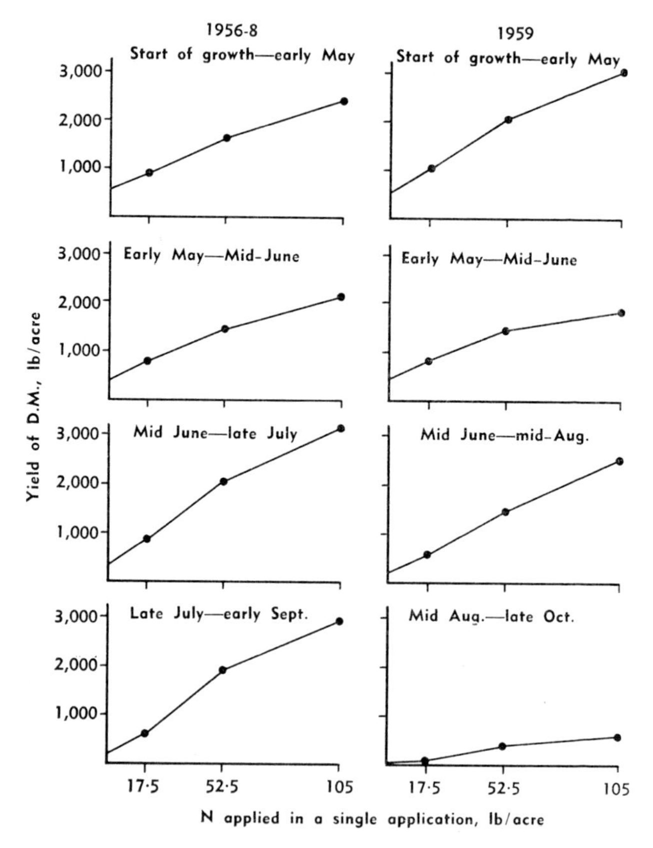

Fig. 14. Yield of DM of grass alone at different levels of N in four successive periods (108).

are shown separately. It is clear that in 1956-8, the greatest responses were obtained in the cuts taken in late July and early September, and that even in 1959 the response obtained at the mid-August cut was approximately the same as that in early May.

In the experiments on permanent pasture carried out by van Burg (73), residual effects were avoided by using separate plots, kept grazed by sheep, for each application date. The responses to N applied in April and June were approximately the same for a growth period of 4–5 weeks, but for a 7-week growth period, response was greatest to N applied in April. The response to N applied in August was smaller than to April or June applications. In a subsequent experiment, applications of 0 or 240 kg N/acre were given to irrigated swards on one of 17 successive dates, and the herbage harvested 31 days later. The yields obtained are shown in Fig. 15. The response to N increased as the date of application progressed from early March to late April, declined during midsummer, increased again for N applied during July and August and then dropped sharply to virtually nil for N applied in late September.

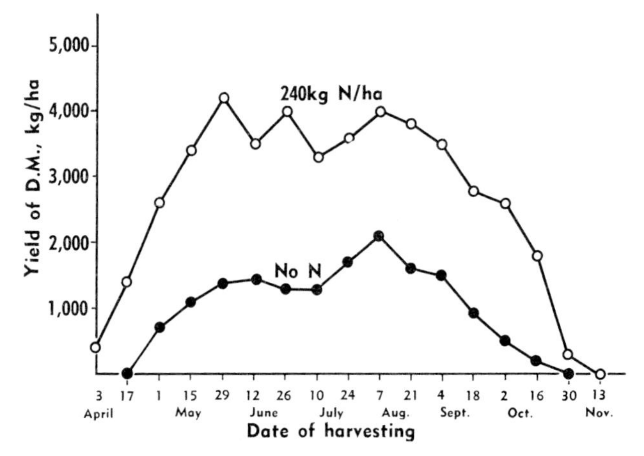

Fig. 15. Yields from irrigated permanent pasture plots treated with nil or 240 kg N/ha on 17 successive dates and harvested 31 days thereafter (73).

Other work carried out on established (irrigated) grass swards in the Netherlands by Sibma (421 and personal communication) showed a continuous decline from April to October in the efficiency of N utilization, i.e. DM production per unit of fertilizer N.

The response of an irrigated S23 ryegrass sward over 4 growth periods was examined in an experiment at Hurley in 1966 (113). Different plots were used for each period. No N was applied until the beginning of the test period, but all the plots were mown on the same dates. The yields obtained for the 8 levels of N are shown in Fig. 16. N response was greatest

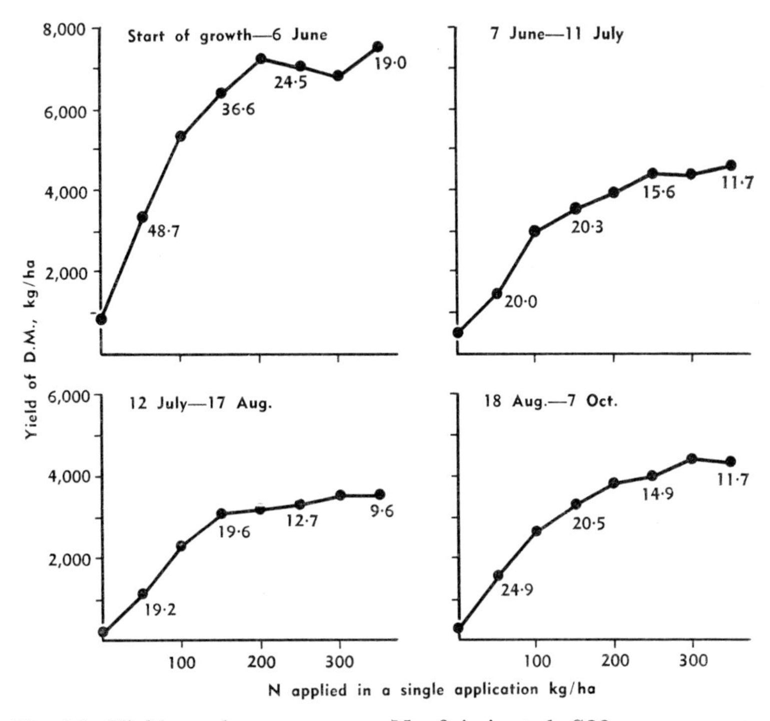

Fig. 16. Yields and responses to N of irrigated S23 ryegrass at 4 periods of the year. Response values in terms of lb DM per lb N are indicated (113).

in the first period, but there was little difference between the other 3 periods in response to the 130-lb level of N.

All the experiments described, with the exception of the 1956–8 portion of the earlier investigation by Cowling (108), showed maximum response to N during May and/or early June, followed by a decline immediately thereafter. The experiments differed, however, in whether or not the response increased during mid-July to early September. The large response in May, followed by a decline in midsummer, is probably due mainly to the change in the physiological condition of the plants. Most studies of growth curves of grasses have shown maximum rates shortly before inflorescence emergence (16), and one would therefore expect response to N to be greatest at this time. After cutting, regrowth requires the formation of new tillers and their development seems likely to be slower the more mature the grass when the first cut is taken. Other differences between seasons and between the results of various experiments are likely to be due mainly to variations in light intensity, temperature, water supply and frequency of defoliation.

Investigations into the effects of the timing of fertilizer N applications on the distribution of herbage DM production through the season have been reported by Castle *et al.* (95). For the first growth period in spring, the optimum timing of N application is influenced by weather conditions, soil type and the form of N used. With winter applications there is a risk of losses through leaching. On the basis of 14 experiments carried out in various parts of England, Devine and Holmes (135) reported that average herbage yields and N recovery were lower from early winter applications than from those in late winter and spring, and that winter applications declined in effectiveness as winter rainfall increased. A similar conclusion was reached by van Burg (76) from trials carried out in the Netherlands during 1959-65.

On the basis of experiments with Italian ryegrass during 7 years in E. Scotland, Heddle (213) reported that applications of fertilizer N in February generally produced yields equal to those from applications made in early March. Applications later than a critical date, which varied from site to site and year to year, but was usually in mid-March, resulted in reductions in yield. The comparisons were based on constant harvesting dates in April or May. In earlier work in S.W. Scotland reported by Hunt (241), fertilizer N applied on 22 March to a sward containing several ryegrass varieties and a little clover was less effective than that applied on 8 March or 22 February. The results of van Burg (76) indicated that the greater the quantity of N to be applied, the earlier the application date could be. He also pointed out that a sward intended for grazing should receive fertilizer N earlier than a sward intended for silage or hay.

Oostendorp (369), working in the Netherlands, recommended that, in practice, N should not be applied before 20 February;

that it should be applied only if the mean minimum temperature had remained above 0°C for 10 days, and should be delayed until the first 10 days of April if the February rainfall exceeded 100 mm, which is more than twice the normal rainfall (76). In 6 out of 7 years, these criteria were reasonably consistent with the results obtained in the trials reported by van Burg (76). The results of these trials also agreed well with the recommendation of Jagtenberg (251) that fertilizer N should not be applied until the accumulated mean daily temperature since 1 January reached 200°C.

Season of the year: experiments in Australia

As stated above, the most marked effects of fertilizer N on grass swards in parts of Australia occur during the winter.

N applied in late autumn (May) produced a linear response of about 28 lb DM/lb N at harvest in August or September for application rates up to 92 lb N/acre on swards predominantly of perennial ryegrass, but with some white clover or subterranean clover, at several sites in southern Victoria (352). In two of the three trials in which a 138-lb N/acre rate was also tested, the response to the additional increment was much less than to 92 lb N/acre.

Other more detailed studies of the N response of swards consisting mainly of Wimmera ryegrass and subterranean clover in Victoria were carried out by Collins (101) over a two-year period. He found that 92 lb N/acre gave yield responses of 8–12 lb DM/lb N for a growth period of 6 weeks after application, and responses of 13–18 lb DM/lb N from two harvests taken over a 12-week period. The additional yield response from applying 184 lb N compared with 92 lb averaged only 4·7 lb DM/lb extra N. Investigations in a phytotron suggested that the limited responses to N in the field in winter were a result of the combined effects of low temperatures and short daylengths.

Length of growth period

The effect of the length of time between the application of N and harvest on yield response is clearly shown by experiments with Italian ryegrass carried out by Wilman (498) in S.E. England. Three levels of Nitro-chalk were applied in March or April and the mean yields obtained over 3 years (total of 8 replicates) are shown in Fig. 17. There were no zero-N plots in these experiments. In each case, temperatures were sufficiently high to allow active growth. Yield increased fairly slowly in the first two weeks and was little affected by the level of applied N. In the last 4 weeks the yield increased at a faster

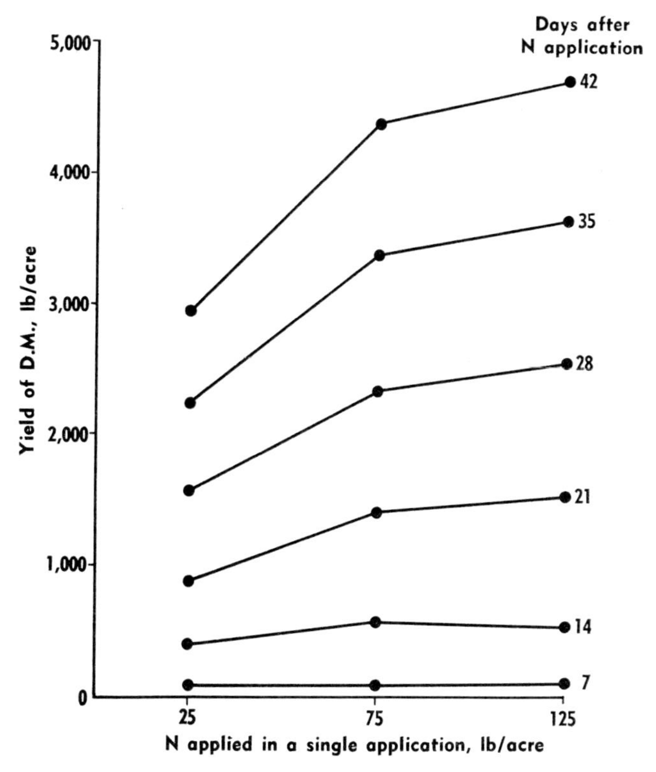

Fig. 17. Yields of Italian ryegrass at weekly intervals following the application of 3 levels of fertilizer N (498).

and fairly constant rate, which was very much influenced by the level of N. The difference at 6 weeks between the 75- and 25-lb rates of N application represented a response of 29 lb DM/lb extra N, but that between the 125 and 75 lb rates was only 7 lb DM/lb extra N. Wilman, however, pointed out that the response to the highest level of N might well have been greater if the crop had been allowed to grow for an extra 2–3 weeks.

A rather similar investigation has been reported from Scotland by Hunt (242), with the results shown in Fig. 18. Up to 16 days after the application of N on 16 April, N rate had little effect. By the 23rd day there was a significant response to 104 lb N, compared with 52 lb, amounting to 10 lb DM/lb additional N. This response increased to 24·4 lb DM/lb additional N by the 57th day (12 June). The highest rate of N application, 156 lb N in a single application, gave no additional response over the 104-lb rate during the 57-day period.

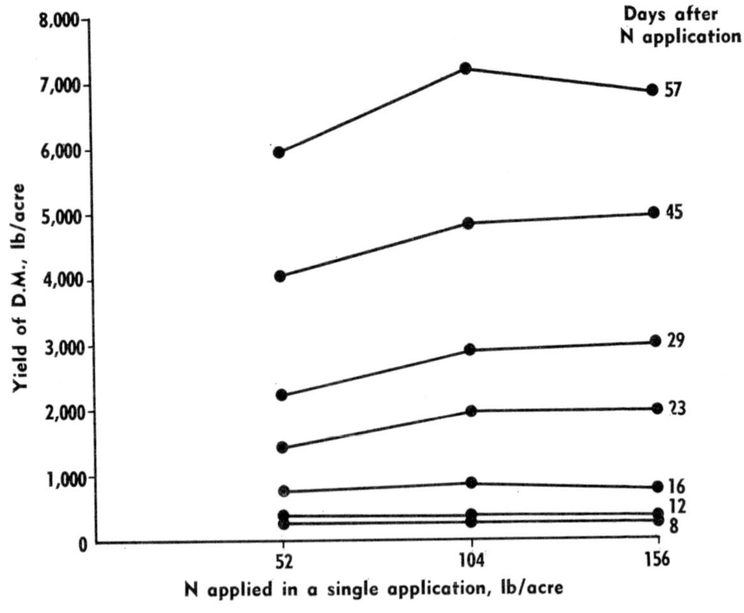

Fig. 18. Yields of Italian ryegrass at intervals following the application of fertilizer N (242).

Hunt also included data on the yield of N in the herbage which, in conjunction with the DM yield data, clearly illustrates the influence of length of growth period on response. The plots cut after 8–45 days were cut again at the end of the 57-day period. Fig. 19 shows the results for the 104-lb N application.

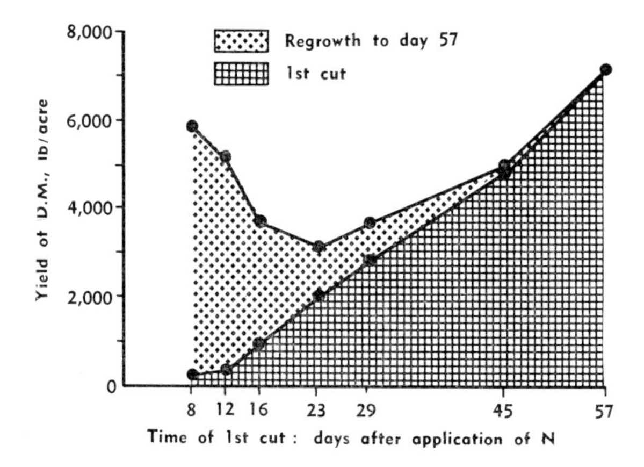

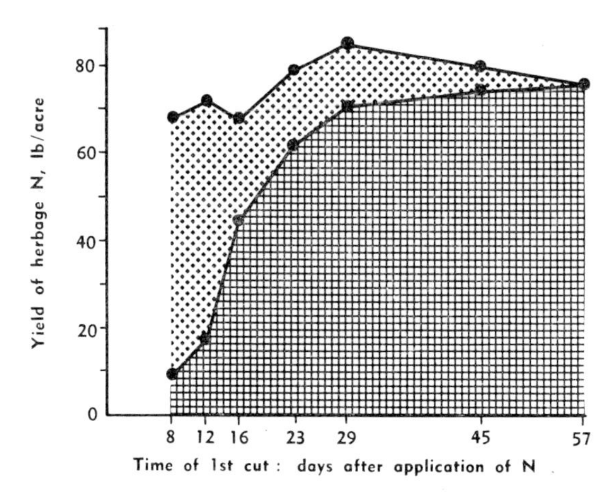

Fig. 19. The accumulation of DM yield and herbage N following the application of 104 lb N/acre (data from Hunt (242)).

When cut 23 days after N application, only 27% of the DM yield obtained after 57 days had been produced, but 82% of the 57-day N uptake had been attained. The response to N in terms of the combined DM yield from the 1st cut and the regrowth was therefore lowest when the plots were cut at this stage.

The results described in this chapter clearly demonstrate that the most efficient use of N is achieved (a) by avoiding the application of quantities excessive in relation to the proposed growth period, and (b) by applying the N for a given growth period (except the first) immediately after the previous harvest.

CHAPTER 26

EFFECTS OF FERTILIZER NITROGEN ON HARDINESS AND DISEASE RESISTANCE

The application of fertilizer N, especially at high levels, may reduce the resistance of herbage plants to adverse weather conditions and to attack by pathogens.

An increased susceptibility to drought, cold and heat, induced in several grass species by ammonium sulphate applied at 137 lb N/acre per year, was demonstrated by Carroll (89). In this experiment cores were removed from both fertilized and unfertilized swards and then subjected to controlled-environment conditions.

The effect of levels of applied N above about 300 lb/acre per year in increasing susceptibility to winter-killing was noted during the severe winter of 1962-3 at Aberystwyth (329), at Hurley (24) and at Rothamsted (494). At Rothamsted, this effect was associated with increased attack by snow-mould fungi (species unidentified). From Georgia, USA, Adams and Twersky (3) reported that winter injury to Coastal Bermudagrass increased at high levels of fertilizer N, but that injury decreased as K supply increased. A similar result was obtained by Kresge and Decker (274). Toomre (452), working in Estonia, however, reported that fertilizer N increased the winterhardiness of grasses, particularly perennial ryegrass.

It is well known that high levels of fertilizer N increase the susceptibility of many crop plants to attack by pathogenic organisms and there is some evidence of this effect in grasses despite their generally high resistance. The incidence of *Helminthosporium* leaf-spot disease on Midland Bermudagrass was greatest on plots receiving high levels of N (400–600 lb/acre per year) with little or no K (274). At Aberystwyth, the incidence of infection of grass by fungi was increased 50% by the application of an unspecified rate of fertilizer N, although yield losses due to infection were outweighed by the increased growth (88).

Goss and Gould (182) found that the incidence of attack by *Ophiobolus* on *Agrostis* was greatest on plots receiving the highest of 3 levels of N in one year but not in another. They suggested that any increase in the susceptibility of individual roots to infection may have been offset by increased total root production. However, this suggestion that root growth is increased by N is not supported by the work summarized on pp. 63–5.

CHAPTER 27

DIAGNOSIS OF NITROGEN DEFICIENCY IN SWARDS

It is unusual for a soil to be able to supply sufficient N for the maximum growth of an all-grass sward. Hence, fertilizer N will almost always produce a yield response. With grass/clover swards the extent to which the grass component is deficient in N depends mainly on the vigour of clover growth. The influence of the presence of clover on the response of a sward to N has been dealt with in Chapter 17.

In practice, the application of fertilizer N to grassland is usually decided on the basis of the performance expected from it. Thus in Britain, the recommendations of the National Agricultural Advisory Service (349) for grazed swards state that N applications should be related to the type of stock, system of management, output required and, under extensive systems, the clover content of the sward, but no mention is made of soil N status. Although the soil contribution is often small in relation to the quantity of fertilizer N required for maximum economic yield, the difference in annual supply between soils previously under arable cultivation and soils that have been under grass for long periods may be more than 100 lb N/acre (see Chapter 9).

Methods based on both chemical extraction and incubation techniques have been proposed for assessing the N status of arable soils, and their relative merits have been reviewed by Bremner (54), Keeney and Bremner (265) and Jenkinson (257). Although results using such methods have sometimes shown good correlations with the growth of newly sown grass in pot trials, there appear to be no published reports of comparisons with the responses of established grass swards to fertilizer N. There is certainly likely to be a considerable difference between the quantity of N released to newly sown grass in the disturbed soil of a pot trial and that released to an established sward. A further difficulty in applying the results of soil analysis to grass swards is that, in nearly all practical situations, complications would arise from the effects of recent fertilizer application or the presence of legumes in the sward.

The various shortcomings of soil analysis for estimating the contribution of soil N to grass swards suggest that herbage analysis might provide a more useful estimate of the N status of a sward. Information is certainly available more quickly from herbage analysis than from soil-incubation techniques, and investigations have been reported which suggest that fairly precise critical levels can be obtained. However, the problem of assessing the fertilizer requirement still remains.

There is evidence that when the N supply of a sward is provided mainly by nitrate N, a nitrate content of 100 m.eq./kg herbage DM, i.e. 0·14% NO_3–N, indicates an adequate supply (75, 504). This value appears to apply to growth periods of 4–6 weeks, and to N supplied as Nitro-chalk or calcium nitrate but not as ammonium sulphate (75). In various trials by van Burg (75), the quantity of fertilizer N required to obtain the critical nitrate value in spring and early summer varied between 90 and 240 kg/ha for grass to be utilized at the grazing stage, but no explanation could be given for the variation in the quantities required in the different trials. In studies in which Italian ryegrass was grown in nutrient culture solutions, Hylton *et al.* (246) reported that the critical level for maximum growth was 0·10% NO_3–N in the youngest fully open leaf with a ligule, a rather lower value than that reported by van Burg.

When N is supplied mainly in the ammonium form, an organic N content greater than 3·5% in the herbage DM indicates that grass in the vegetative stage has an adequate supply (504).

Greenwood *et al.* (196) have recently pointed out that while herbage nitrate contents may be useful for indicating when N supply is adequate for maximum yield, they are unsatisfactory for indicating the magnitude of a deficiency. They have therefore investigated other means of assessing the intensity of deficiency, which they term ‘ nitrogen stress ’. Using *Lolium rigidum* they found close relationships between N stress and (a) concentration of total N, (b) concentration of free ninhydrin-positive N, and (c) leaf elongation, but the relationships differed at different stages of development. Further work indicated calibration values for these 3 indices for various stages of growth up to 9 weeks after seedling emergence (197).

Despite the work that has been carried out on soil and plant analysis, it seems probable that the rate of fertilizer N application to grassland will continue to be decided mainly on the empirical basis of sward management and yield required, a procedure which normally produces satisfactory results. Herbage and soil analyses, however, may well provide useful information on the N status of swards relying on soil and/or clover N. Herbage analysis may also be used to assess the extent to which a given application of fertilizer N meets the requirement for maximum growth in a particular situation.

CHAPTER 28

RECOVERY OF FERTILIZER NITROGEN AND ITS REMOVAL IN CUT HERBAGE

In calculating the percentage recovery of fertilizer N, i.e. the proportion of that added which is recovered in the harvested herbage, allowance is made for the N contributed by the soil by subtracting an amount equivalent to that in the herbage from comparable plots receiving no fertilizer N. Some error may be involved in this calculation, since the fertilizer application may itself influence the release of available N from the soil.

The recovery of fertilizer N by grass is normally greater than that by arable crops, often amounting to 55–70% of that applied, depending on management and other factors affecting yield response. At Rothamsted, recovery values have often exceeded 70% (103). Apparent recovery values have occasionally exceeded 100%, suggesting that in these instances fertilizer application has increased the availability of soil N.

With rates of N up to about 400–500 lb/acre per year, percentage recovery is generally highest at rates of 200–300 lb. With light applications, much of the N is likely to remain in the roots and stubble, and with high rates of N a relatively large proportion is likely to be unabsorbed. This trend is illustrated by the results reported in Table 10 for Italian ryegrass cut 4 times per year.

TABLE 10

Herbage N yield and % recovery of fertilizer N by Italian ryegrass
(**Data from I.C.I. Ltd (249)**)

	N applied, lb/acre/year								
	0	94	188	282	376	470	564	659	753
N yield in herbage, lb/acre/year	61	105	171	237	275	308	343	357	374
Applied N recovered, %	—	47	59	62	57	53	50	45	42

The mean values for herbage yield and percentage recovery of fertilizer N obtained by Reid and Castle (391) for S23 perennial ryegrass and S37 cocksfoot cut 6 times per year in 3 successive years are given in Table 11.

TABLE 11

Herbage N yield and, in brackets, % recovery of fertilizer N; mean values for ryegrass and cocksfoot. (**Data from Reid and Castle** (**391**))

	N applied, lb/acre/year			
	0 lb N/acre	104 lb N/acre	208 lb N/acre	312 lb N/acre
1960	59	107 (46)	176 (56)	240 (58)
1961	37	99 (60)	178 (68)	243 (66)
1962	29	93 (62)	171 (68)	240 (68)

Grass species differ in their ability to recover fertilizer N. Results of trials carried out by Cowling and Lockyer (116) are given in Table 12 and show that of the species tested, N recovery was highest for cocksfoot.

TABLE 12

Herbage N yield and, in brackets, % recovery of fertilizer N of various grass varieties; mean of 3 years. (**Data from Cowling and Lockyer** (**116**))

	N applied, lb/acre/year			
	0 lb N/acre	81 lb N/acre	163 lb N/acre	326 lb N/acre
Perennial ryegrass, S23	11	43 (39)	90 (48)	218 (63)
Perennial ryegrass, S24	19	48 (35)	98 (48)	218 (61)
Perennial ryegrass, Irish	18	48 (38)	94 (46)	180 (50)
Cocksfoot, S37	23	71 (59)	125 (63)	251 (70)
Timothy, S48	24	63 (48)	117 (58)	229 (63)
Meadow fescue, S215	19	55 (44)	107 (54)	228 (64)

A higher recovery of fertilizer N by cocksfoot than by meadow fescue, timothy and perennial ryegrass was also reported by Widdowson *et al.* (490). Cocksfoot, however, made least use of soil N.

The recovery of N from a single application is influenced by the time-interval between application and harvest, as shown by the results of Wilman (498), given in Table 13. In this experiment with Italian ryegrass, N was applied at 25, 75 and 125 lb/acre. As there were no zero-N plots, the recovery figures given below refer to the 50-lb difference between the 25- and 75-lb rates and the 100-lb difference between the 25-and 125-lb rates.

TABLE 13

Percentage recovery of fertilizer N by Italian ryegrass at various intervals after application. (Data from Wilman (498))

N applied (in addition to lowest rate of 25 lb)	Weeks after application					
	1	2	3	4	5	6
	% Recovery					
50 lb	2	34	70	78	86	76
100 lb	1	20	59	67	72	74

In comparisons between ammonium nitrate, ammonium sulphate and urea at 10 sites in England and Scotland, Devine and Holmes (133) found that N recovery varied considerably from site to site, but that with an annual rate of 180 lb N/acre, recovery at each site was lowest from urea. Recoveries of N from ammonium nitrate and ammonium sulphate were similar except at one site with a soil pH of 8·2, where recovery from ammonium sulphate was low.

Very low recovery values (45% or less) obtained with normal growth periods of 3–6 weeks and a cutting height of 1–2 in. are likely to be due to limitations on herbage growth imposed by factors other than N supply. With the more usual recovery values of 50–80%, some of the N not accounted for is likely to have been absorbed by the grass and to have remained in the roots and stubble, which will then have contained more N than the roots and stubble of the control plots not receiving fertilizer N. Net immobilization of N in the soil OM (other than roots) does not appear to be significant. Other possible factors involved are leaching (see Chapter 11) and gaseous losses (see Chapter 12).

PART III

EFFECTS OF FERTILIZER NITROGEN ON THE COMPOSITION AND QUALITY OF HERBAGE

CHAPTER 29

EFFECTS ON THE BOTANICAL AND CHEMICAL COMPOSITION OF MIXED SWARDS

A major effect of the application of fertilizer N to grass/clover swards is to suppress the clover component (see Chapter 3). Because grasses differ considerably from clovers in chemical composition, any change in the grass/clover ratio will also change the overall chemical composition of the herbage.

White clover is more digestible than grass, particularly when both are fairly mature, and its digestibility declines more slowly over the spring and early summer period (129, 207). Thus the suppression of clover by fertilizer N will also reduce the overall digestibility of the herbage, the extent depending on the time of harvesting.

Clover herbage also generally has a higher N content than grass, even when the grass is given moderate to high levels of fertilizer N. Herbage N contents in white clover are usually within the range 3·5–5·0%, while those in red clover and lucerne are rather lower (485).

However, since the application of fertilizer N tends to increase the N content of grass (see p. 133), the net effect on the herbage of grass/clover swards is variable. In their trials on mixed swards at 6 centres in Scotland, Reith *et al.* (396) found that, in some instances, percentage contents of N in herbage which had received 174 lb N/acre per year were lower than in that not receiving fertilizer N, but that herbage from swards receiving 348 lb N always had higher N contents than that from swards given no N. Reid and Castle (391) reported that in trials during 3 years, the herbage N content of ryegrass/white clover and cocksfoot/white clover swards cut 6 times per year was reduced from an average of 2·66% with no fertilizer N, to 2·41% with 104 lb N and to 2·46% with 208 lb N/acre per year. Where 312 lb N/acre was applied, the percentage N in the herbage was depressed in the 1st year, but in the other two years was higher than in control plots not receiving N. Cowling and Lockyer (116) noted that herbage from grass/clover swards given no N and cut 4 or 5 times per year had higher N percentages than the same grasses each grown alone and given up to 348 lb N/acre per year.

As well as having a higher total N content, clovers usually have a higher proportion of their N in organic form, and usually not more than 1% of their N is present as nitrate (429). In contrast, up to 10% of the N in grass receiving high levels of fertilizer N may be in the form of nitrate, even when cuts are taken at intervals of 4–6 weeks (see pp. 135–7).

Grasses and clovers also differ in their contents of mineral elements. Clovers generally have considerably higher contents of Ca and slightly higher contents of Mg in the early part of the growing season (see review by Whitehead (482)) and this difference may be important for livestock nutrition in some situations. However, with most other nutrient elements, differences between grasses and clovers are not consistent. Thus, clovers show a much wider variation in Cu content than grasses; when Cu is plentiful, clovers tend to have higher contents than grasses, but when in short supply, grasses usually have higher contents than clovers (482). Co contents, however, are generally higher in clovers than in grasses.

Reports have indicated variable effects of fertilizer N on the balance between various grass species but, in general, the more productive species appear to be encouraged by additional N. Thus Holmes (233) reported that Nitro-chalk at 200–300 lb N/acre per year increased the desirable species perennial ryegrass and cocksfoot relative to the less desirable *Agrostis*, *Holcus* and

Poa spp. Reith *et al.* (396) also found that in most of their trials, ' other grasses ' (mainly *Poa* spp.) decreased where the highest rate of Nitro-chalk (348 lb N/acre per year) was applied. In contrast, McAllister and McConaghy (305) noted that *Poa* spp. tended to become dominant in plots receiving a high level of Nitro-chalk (367 lb N/acre per year). On a mixed sward, plots receiving 200 lb N/acre per year had higher contents of *Poa annua* and perennial ryegrass and lower contents of *Agrostis* (and white clover) than plots receiving no fertilizer N (68). Raininko (385) reported that fertilizer N at rates of 100 and 200 kg N/ha increased the proportion of cocksfoot in mixtures with timothy and with meadow fescue.

Observations on permanent pasture in the Netherlands indicated that increasing the rate of applied N increased both perennial ryegrass and *Poa* spp. at the expense of cocksfoot and *Agrostis* (209). On permanent pasture in E. Germany, Kreil *et al.* (273) found that increasing the rate of fertilizer N up to 720 kg N/ha per year increased the proportions of *Poa pratensis* and *Agropyron repens* and, at levels above 240 kg/ha, decreased perennial ryegrass. Ennik (157) reported little difference in the botanical composition of permanent pasture receiving either 70 or 140 kg N/ha per year.

The influence of N supply on the balance between species is likely to be modified by such factors as defoliation frequency and the supply of other nutrients. Thus, in a 12-year experiment, Castle and Holmes (93) found that with N-fertilized plots, those which had also received K had substantially higher proportions of perennial ryegrass, meadow fescue and cocksfoot than plots which had not received K.

Reports of the influence of fertilizer N on the incidence of dicotyledonous non-legume species also lack agreement. Increasing levels of fertilizer N have been reported to cause decreases (209, 305), increases (446), and to have little effect (273).

CHAPTER 30

EFFECTS ON CONTENTS OF DRY MATTER AND OF NON-NITROGENOUS ORGANIC CONSTITUENTS

Dry matter

Increasing the level of fertilizer N tends to decrease the DM content of herbage. The extent of this is impossible to assess precisely, since the heavier crop resulting from the application of N holds a greater amount of superficial water. However, Lazenby and Rogers (285) found that applications of fertilizer N to ryegrass over the range 0–800 lb/acre per year decreased the average DM content from 23 to 17% of the fresh weight. De Groot and Keuning (204) reported that mean DM contents over a 5-year period for plots receiving 150 and 600–650 kg N/ha per year were 15·5 and 13·7%, respectively.

Fibre and carbohydrates

The fibre content of herbage DM is little influenced by fertilizer N (204, 386, 394, 505). Its component materials cellulose (50, 71) and lignin (50, 394) are also little affected. Reid *et al.* (393), however, did find a small reduction in the cellulose and acid-insoluble-lignin content of cocksfoot with levels of fertilizer N up to 504 kg/ha, and Bezeau *et al.* (40) reported that the cellulose content of *Festuca scabrella* decreased with increasing level of fertilizer N.

The content of water-soluble carbohydrate is often markedly decreased by the application of fertilizer N (50, 71, 181, 259, 260, 362, 388, 389, 394). With perennial ryegrass, Jones *et al.* (260) obtained the following contents of water-soluble carbohydrate in samples taken on 14 May following a single application of Nitro-chalk at the end of February :

N applied, lb/acre	17·5	35	70	140
Water-soluble carbohydrate, % of DM	35·3	32·9	24·1	19·1

The average values obtained in a 1-year experiment by Reid (389) for 5 cuts from a perennial ryegrass/timothy sward given a wide range of N rates are given below :

N applied, lb/acre/year	0	100	200	300	400	500	600	700	800
Water-soluble carbohydrate, % of DM	14·6	14·9	13·0	12·2	10·6	10·3	8·8	8·7	8·2

In experiments in Virginia, Reid *et al.* (394) obtained the following values for cocksfoot herbage :

N applied, lb/acre (single dressing)	0	50	100	400
Water-soluble carbohydrate, % of DM	10·0	9·3	5·0	4·5

Part of the difference between these 3 sets of values is probably attributable to differences between species, cocksfoot having lower contents of soluble carbohydrate than the other species, and to differences in the number of cuts and time of cutting (466).

Smaller effects were found by Rhykerd *et al.* (398), working in Indiana, who reported that the application of up to 1200 lb N/acre per year consistently reduced the average water-soluble-carbohydrate content of cocksfoot from 5·66 to 4·01%, but did not significantly influence the content in smooth bromegrass or timothy. Small and inconsistent effects on soluble-carbohydrate content in cocksfoot were also found by Reid *et al.* (393).

The change in water-soluble-carbohydrate content is proportionately much greater in the polysaccharide fraction than in the mono- and di-saccharides (71, 362).

When grass is consumed as such by livestock, the depression in soluble carbohydrate caused by N application is thought to be of little importance, since overall OM digestibility is only slightly affected (388), the decrease in soluble carbohydrate being accompanied by an increase in soluble N-containing constituents. However, the production of good silage does require the presence of adequate soluble carbohydrate (466).

Fats and fatty acids

Ramage *et al.* (386) reported that rates of up to 400 lb N/acre per year had no effect on the ether-extract content of the herbage of cocksfoot or reed canarygrass. Although Immink *et al.* (247) demonstrated a correlation between the contents of higher fatty acids and total N in grass herbage from permanent pasture, no mention is made of fertilizer N rates.

Organic acids

Blackman and Templeman (46) reported that the application of calcium nitrate and, to a lesser extent, ammonium sulphate generally increased the percentage of total organic acids in grasses. A correlation between the organic anion content of ryegrass and both total N and nitrate N contents has been reported by van Tuil (454). Cocksfoot receiving 400 lb N/acre contained 80% more malate than that receiving 30 lb N (118).

Vitamins and carotene

The application of fertilizer N has been shown to increase the carotene content of herbage by up to 40% (340, 427). Döring (146) reported that increasing rates of N considerably increased the concentration of vitamin A (presumably as carotene) and vitamin B in ryegrass. Concentrations of vitamins C, D and E showed some increase with moderate levels of N, but decreased with higher levels. Carey *et al.* (87) reported that fertilizer N increased the content of riboflavin in bromegrass.

Alkaloids

There is very little information on the effect of fertilizer N on the alkaloid content of herbage. However, Bennett (37) has reported that high levels of nitrate N given to ryegrass growing in a loam/sand mixture in pots, increased the content of perloline up to 20-fold.

CHAPTER 31

EFFECTS ON CONTENTS OF TOTAL NITROGEN AND MAJOR NITROGEN-CONTAINING CONSTITUENTS

Increases in the rate of fertilizer N to grass generally result in a progressive increase in the N content of the herbage (116, 249, 389, 398). However, relatively small amounts of N, although increasing DM yield, may sometimes decrease herbage N contents (see Fig. 20 and Cowling and Lockyer (116)). High rates of N continue to produce increases in herbage N content even after the ceiling for maximum DM yield has been reached.

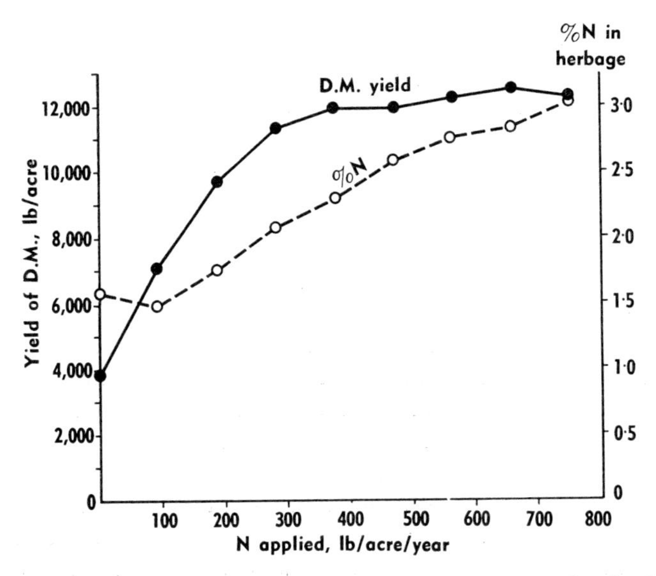

Fig. 20. The influence of increasing N supply on the herbage yield and N % of Italian ryegrass cut four times per year (249).

The effect of N supply on herbage N content is influenced by the time-interval between N application and sampling. This is clearly shown by the data of Wilman (498) (see Fig. 21) obtained from trials with Italian ryegrass harvested weekly over 6-week periods following N application.

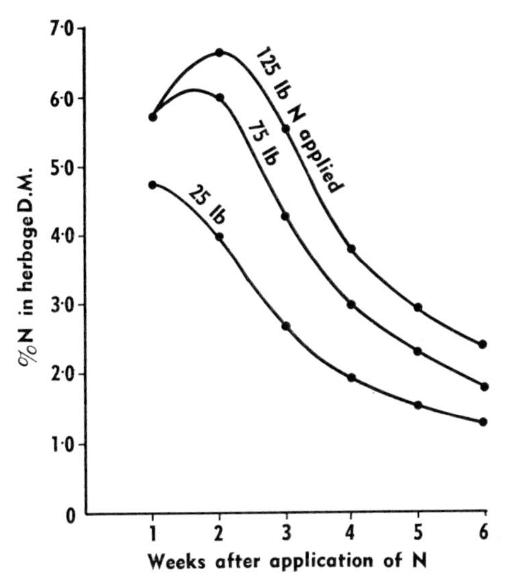

Fig. 21. The influence of increasing N supply on the herbage N % of Italian ryegrass sampled at weekly intervals for 6 weeks after the application of fertilizer N (498).

A substantial proportion, often 15–25%, of the total N in both grass and legume herbage exists in various non-protein forms. Much of this non-protein N is organic and consists of amino acids, amides and peptides, with smaller amounts of a wide range of compounds such as purines, betaine and choline (161, 162). Of the inorganic forms, ammonium N usually occurs in trace amounts only. Nowakowski *et al.* (365) found less than 2% of the total N in this form in Italian ryegrass, even in the presence of very high levels of ammonium in the soil, and less than 0·7% of the total N as ammonium when N was supplied as nitrate. Only 0·5% of the total N occurred in the ammonium form in bromegrass that had received 300 lb N/acre as ammonium nitrate (87). However, the nitrate content of herbage can vary from a trace to more than 50% of the total N in exceptional circumstances. Nowakowski *et al.* (365) reported that Italian ryegrass grown under conditions of high nitrate supply in a pot experiment contained 2·79% nitrate N, representing 51% of the total N.

There appears to be little information from field experiments on the effects of fertilizer N on the contents of organic non-protein N. However, data for Italian ryegrass grown in a glasshouse at various levels of ammonium N and nitrate N supply under controlled conditions of light and temperature have been reported by Nowakowski *et al.* (365) and Nowakowski and Cunningham (364). Their results showed that the proportion of total N in soluble organic non-protein form was much greater when no N was applied or when N was supplied as ammonium rather than as nitrate. Increasing the supply of ammonium up to 400 ppm. had little effect on the proportion of organic non-protein N in the first experiment (365), but approximately doubled this fraction in the second (364). Grass given ammonium had more soluble organic N when the light intensity was decreased. Carey *et al.* (87) found that an application of 300 lb N as ammonium nitrate to bromegrass produced a peak amide-N content of 0·07% of the DM (equivalent to about 2% of the total N) compared with about 0·01% of the DM in unfertilized grass. Soluble organic N is also increased by K deficiency (363). In particular, amides such as asparagine tend to accumulate with K deficiency (201).

As indicated above, nitrate is the only form of inorganic N which accumulates to any appreciable extent in herbage, and the dominant factor influencing this is the supply of nitrate to the plant roots. Nowakowski and Cunningham (364) showed that when N was applied as ammonium and the nitrification inhibitor 'N-Serve' was added in pot trials with Italian ryegrass, high N rates had little effect on the nitrate content of herbage. However, in practice, N is most commonly applied in the form of ammonium nitrate, and some of the ammonium is nitrified before uptake. Recent data showing the increase in nitrate content in perennial ryegrass/timothy herbage with increasing levels of N applied as Nitro-chalk have been provided by Reid (389). With rates between 400 and 800 lb N/acre per year, most of the additional N in the herbage was in the form of nitrate. Table 14 gives the mean values for nitrate N contents from 5 cuts following N applied in equal quantities in early April and then after each cut except the last.

TABLE 14

Accumulation of nitrate in ryegrass/timothy herbage with increasing rates of fertilizer N. (**Data from Reid** (**389**))

	N applied, lb/acre/year								
	0	100	200	300	400	500	600	700	800
Total N, % in DM	2·27	2·40	2·59	2·85	3·30	3·20	3·54	3·54	3·74
Nitrate N, % in DM	·011	·019	·035	·084	·254	·313	·408	·492	·531
Nitrate N as % of total N	0·48	0·79	1·35	2·95	7·70	9·78	11·5	13·9	14·2

Kershaw (268) showed that calcium nitrate generally produced higher nitrate contents in Italian ryegrass than did ammonium sulphate, but that the differences were small. The application of a nitrification inhibitor decreased the nitrate content of grass supplied with ammonium sulphate (366).

The actual contents of nitrate N present in herbage are greatly influenced by the time-interval between the application of fertilizer and sampling. This is clearly illustrated in Fig. 22, which is based on Wilman's data (498).

Weather and nutritional factors can also have a considerable effect. It seems probable that any factor which restricts growth rate or protein synthesis without restricting nitrate uptake to an equivalent extent, will tend to increase the nitrate content in herbage. Certainly, nitrate contents are appreciably increased by low light intensities (5, 34, 130, 364).

Nowakowski *et al.* (365) found that increasing the soil temperature over the range 11–28°C substantially increased nitrate contents in Italian ryegrass at each of 6 levels of N application. Bathurst and Mitchell (34) also reported that the nitrate-N content of various grasses, grown under controlled conditions, tended to increase as the temperature was increased from 7 to 24°C.

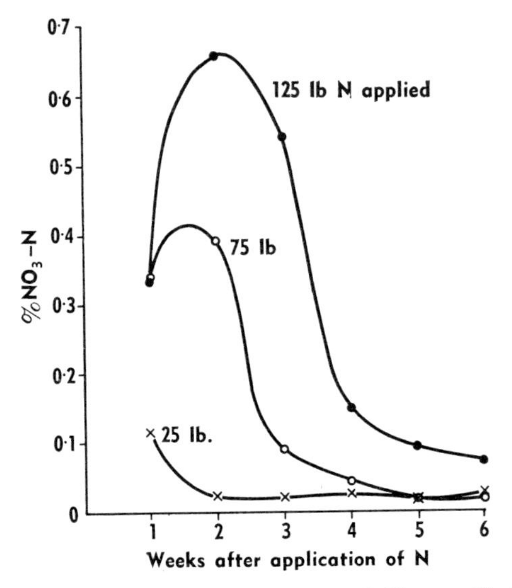

Fig. 22. The influence of three levels of N, applied as Nitro-chalk, on the content of nitrate-N in Italian ryegrass sampled at weekly intervals for 6 weeks after application (498).

S deficiency has been reported to increase the proportion of nitrate N in grass herbage (1, 31), an effect probably explained by the need for S in protein synthesis. The metabolism of nitrate also requires the participation of enzymes containing Mn, Mo, Fe and Cu, and deficiencies of these elements may accentuate nitrate accumulation (81, 514).

CHAPTER 32

EFFECTS ON CONTENTS OF MINERAL ELEMENTS

The effects of fertilizer N on percentage contents of other elements in herbage as reported in the literature have been inconsistent.

Increases in percentage contents of P, K, and possibly of other elements also, have been brought about by N applications when these elements were abundant, and decreases when they were not. There are differences in the effects caused by the various forms of N, particularly for those elements (Fe, Mn, Co) whose uptake is most influenced by pH. Thus the use of ammonium sulphate, by tending to increase soil acidity, results in greater uptake of Fe and Mn than does Nitro-chalk. The effects of given forms of fertilizer N are also modified by the other elements they contain. Thus, ammonium sulphate produces herbage with higher S contents than Nitro-chalk; and Nitro-chalk produces higher Ca contents than ammonium sulphate at equal rates of N application. The effects of ion competition during uptake by plant roots also differ with the form of fertilizer N. Thus, there is some evidence, as indicated below, that anhydrous ammonia produces herbage with lower contents of K, Ca and Mg than does an equal quantity of N as Nitro-chalk, presumably because the uptake of ammonium depresses the uptake of other cations.

When considering mineral contents in relation to animal requirements, it should be noted that fertilizer application may influence the availability of an element to livestock, as well as its percentage content; also that changes in availability and content may either augment or counteract one another. For example, Hartmans and van der Grift (210) showed that the availability of Cu to cattle was inversely related to the S content of the herbage (which is increased by the use of ammonium sulphate).

Phosphorus

The reported effects of N application on percentage P contents in herbage have been variable. In trials reported by Walker *et al.* (472), N treatments decreased herbage P contents on soils low in P, and increased them on soils rich in P. Stewart and Holmes (437) and Reith *et al.* (396) found that fertilizer N had almost no effect on the herbage P content of grass/clover swards. Several experiments at the Grassland Research Institute, Hurley (483) have shown reductions from about 0·42% to about 0·30% in the P content of grass following the applications of high rates of N to herbage which was cut and removed, but no effects on P content when the herbage was grazed. Kershaw and Banton (269) found that N at rates up to 396 lb/acre increased herbage P content by 50% at the 1st cut, caused a much smaller increase at the 2nd cut and depressed it at the 3rd. MacLeod (315) reported that 600 lb N/acre as ammonium nitrate per year depressed the mean P content of 3 grass species from 0·32 to 0·18%.

Ammonium sulphate and calcium nitrate had similar effects on P content, indicating that sulphate did not compete more strongly than nitrate with P uptake (269). Reid *et al.* (394) found that sodium nitrate, ammonium nitrate, ammonium sulphate and urea produced almost identical P contents in tall fescue herbage.

Potassium

The effect of N on K content is influenced by the supply of available K. Results obtained by Kemp (266) indicated that N applications caused a decrease when herbage K content in control plots not receiving N was below 2%, and an increase when the content was above 2%. In general, the more recent data have supported the view that changes in K content induced by N applications are influenced by K supply, but some experiments have shown decreases when the K content in controls was above 2%. Thus, McConaghy *et al.* (309) reported that 350 lb N/acre per year as Nitro-chalk depressed herbage K contents from 2·25 to 1·65%; and Mortensen *et al.* (342) that 500 lb N as ammonium nitrate reduced the K content of cocksfoot from 3·2 to 2·7%.

There was no significant difference between calcium nitrate

and ammonium sulphate in their effect on the K content of Italian ryegrass (269). Anhydrous ammonia, however, produced herbage with lower K contents than the same amount of N (250 kg/ha) applied as nitro-lime (77). There is also evidence that urea produces lower herbage K contents than nitrate or ammonium salts (394).

Calcium

In many experiments, the use of Nitro-chalk, which contains $CaCO_3$, has made it difficult to establish the effect of N on herbage Ca contents. However, in pot trials, Nielsen and Cunningham (355) found that nitrate N greatly increased the Ca content of Italian ryegrass herbage, while ammonium N reduced it slightly.

In a number of field experiments, moderate levels of Nitro-chalk have decreased the Ca content of herbage, while high levels have increased it (482, 483). Ammonium nitrate at rates up to 504 kg N/ha had little effect on the Ca content of cocksfoot (393).

With anhydrous ammonia and nitro-lime each applied at 250 kg N/ha, van Burg and van Brakel (77) found that the Ca contents of grass herbage from plots treated with ammonia averaged 8% lower on a clay soil and 22% lower on a sandy soil, than those from plots receiving nitro-lime. Urea has also been reported to produce lower herbage Ca contents than nitrate and ammonium salts (394).

Magnesium

In many field experiments, N applications, mostly as Nitro-chalk, have increased herbage Mg contents of all-grass and grass/clover swards (42, 214, 437). However, Gardner *et al.* (167) and Mortensen *et al.* (342) reported ammonium nitrate to have no effect.

Nitrate and ammonium forms of N have different effects on herbage Mg contents when taken up in these forms. Thus, Nielsen and Cunningham (355) found in pot experiments that nitrate, but not ammonium, increased the Mg content of Italian ryegrass. In the field, of course, ammonium is partially converted to nitrate before being absorbed by plants.

Kershaw and Banton (269) found that N applied as ammonium sulphate produced significantly lower Mg levels in Italian ryegrass than N applied as calcium nitrate; and Wolton (510) that ammonium sulphate produced herbage with lower Mg contents than did ammonium nitrate. Herbage from a sward treated with anhydrous ammonia was found by van Burg and van Brakel (77) to contain 13–19% less Mg than that treated with an equal quantity of N as nitro-lime.

Sulphur

The S content of herbage is increased by ammonium sulphate but not by other N fertilizers.

Jones (261) reported that the mean herbage S content of 3 mixed grass/clover swards was increased from 0·27 to 0·45% by the application of 6 cwt ammonium sulphate/acre per year for 3 years, and Conroy (102) found that the application of 16 cwt ammonium sulphate/acre increased the S content of grass/clover herbage from 0·38 to 0·57%, but ammonium nitrate had no effect. Ammonium nitrate also had no effect in an experiment reported by Rahman *et al.* (384).

Sodium

The effect of fertilizer N on the percentage Na content of herbage varies markedly with species.

N applications have been reported to have little effect on the Na content of timothy and meadow fescue, which are inherently low in Na content, whereas with perennial ryegrass, Italian ryegrass and cocksfoot, N has considerably increased Na content (199, 483).

With mixed grass/clover swards, Reith *et al.* (396) found that the mean herbage Na content from 4 sites was increased from 0·39 to 0·64% by 174 lb N/acre per year as Nitro-chalk and to 0·69% with 348 lb N.

Chlorine

Conflicting results have been reported on the effects of fertilizer N on herbage Cl contents.

In a pot experiment, Dijkshoorn (136) found that ammonium nitrate at rates up to about 800 kg N/ha depressed the Cl content of ryegrass by up to 40% when KCl was also added, but had only a slight effect in its absence.

In a field trial, Rahman *et al.* (384) found that 336 lb N/acre per year as ammonium nitrate reduced the Cl content of perennial ryegrass herbage from an average of 0·53 to 0·46% and that of perennial ryegrass/white clover from 0·57 to 0·42%. In a trial with a range of N rates up to 340 lb/acre per year, the Cl content of various grass species was generally increased about 50% by 170 lb N/acre, but to a lesser extent by 340 lb/acre as Nitro-chalk (483).

Iron

Plant uptake of Fe is greatly influenced by soil pH. Fertilizers which tend to increase soil acidity, such as ammonium sulphate, can therefore be expected to increase plant Fe contents. Some evidence of this was found in trials at Hurley (483). Hemingway (215), however, found no increase over a 3-year period in which 12 cwt ammonium sulphate/acre per year was applied.

Manganese

The uptake of Mn by plants is normally highest on acid soils and Hemingway's (215) report that the average Mn content of grass receiving 12 cwt ammonium sulphate/acre per year was 79 ppm., compared with 21 ppm. for grass not receiving N, is consistent with this. Nitro-chalk was found to cause substantial decreases in the Mn content of herbage (397).

Zinc

Some investigations have shown herbage Zn contents to be increased by N rates above about 400 lb/acre per year (331, 483), although no increase occurred in the experiments of Reid *et al.* (394). Lower levels of N appear to have no effect (331, 396, 483).

Copper

Conflicting results have been reported on the effects of fertilizer N on herbage Cu contents.

Hemingway (215) found that ammonium sulphate increased the Cu content of grass but not of clover.

With mixed swards, increases in herbage Cu contents with increasing fertilizer N supply have been reported by Stewart and Holmes (437) and Havre and Dishington (211), but decreases by Knabe *et al.* (272), Bosch (52) and Kreil *et al.* (273).

In experiments at Hurley, the Cu content of grazed ryegrass given 980 lb N/acre per year was about 9 ppm. compared with about 5 ppm. where it was given only 200 lb N/acre. However, on cut swards the application of 280 lb N/acre produced slightly lower Cu contents in grass than no fertilizer N (483).

Cobalt

As with the uptake of Fe and Mn, soil pH is an important factor influencing Co uptake by plants.

Nitro-chalk applied at 350 lb N/acre per year decreased the Co content of mixed herbage from 0·13 ppm. with no N, to 0·08 ppm. (397). No consistent effect of increasing rates of urea up to 448 kg N/ha was found by Reid *et al.* (394).

Iodine

Alderman and Jones (7) reported that monthly applications of 4 cwt ammonium sulphate/acre significantly reduced herbage I contents of 4 grasses compared with the much lower-yielding control plots.

Molybdenum

Mo uptake is reduced by acid conditions, and a reduction in herbage Mo content following the application of ammonium sulphate has been reported (215). Nitro-chalk had no consistent effect in the experiments reported by Reith and Mitchell (397).

Silica

Increasing N supply results in a decrease in the silica content of cereals (262) and a similar effect is to be expected in pasture grasses. However, Reid *et al.* (394) reported no marked effect.

CHAPTER 33

EFFECTS ON PALATABILITY, INTAKE, DIGESTIBILITY AND METABOLIC UTILIZATION OF HERBAGE

Palatability

There are conflicting reports on the effect of fertilizer N on herbage palatability assessed, in most instances, in situations where the animals had free choice between herbage from the various treatments.

Ivins (250) found that herbage from a grass/clover sward receiving 35 lb N/acre as Nitro-chalk was eaten more readily than herbage receiving no N by cows given access to both areas. Burton *et al.* (80) reported that the palatability of Coastal Bermudagrass to cows grazing plots receiving a range of rates of ammonium nitrate was increased by rates up to several hundred pounds of N/acre per year, and that there was no evidence of a decrease in palatability at the highest level of 1500 lb N. On the other hand, Sonneveld (430) observed that sheep and cattle appeared to find grass fertilized with 500 kg N/ha per year less palatable than that receiving only 100 kg N or less. McAllister and McConaghy (305) noted that cattle grazing plots receiving 365 lb N as Nitro-chalk ate the herbage less readily than cattle grazing plots receiving 182 lb N or less; and Reid and Jung (392) found that sheep fed indoors showed a preference for tall fescue hay that had received a low level of N. Marten and Donker (321) reported no difference in the palatability to grazing cattle of bromegrass given nil or 300 lb N/acre per grazing either as ammonium nitrate or as urea.

Reid *et al.* (394) carried out two trials in which the influence of increasing levels of N up to 400 lb/acre as urea on the palatability of cocksfoot to sheep was investigated. When spring-growth herbage was made into hay, its palatability (intake of individual hay sample as percentage of total *ad lib.* intake) declined with increasing level of N. However, in a grazing trial with sheep in the autumn, consumption from the plots receiving 100, 200 and 400 lb N/acre was significantly greater than from plots receiving 0 and 50 lb N. Two factors may have contributed to this difference between the trials. Firstly, in the grazing trial the application of fertilizer increased the quantity of herbage available, thus encouraging consumption from the high-N plots. Secondly, increasing N application in spring consistently decreased the soluble-carbohydrate content, but had little effect on the structural constituents, whereas in autumn the effect on soluble carbohydrate was much less, but there was some depression of cell-wall constituents, including lignin.

The form in which fertilizer N is applied influences herbage palatability. Grass given 386 lb N/acre per year as ammonium sulphate was eaten less readily by cattle grazing over both plots than that receiving the same amount of N as Nitro-chalk (339). At rates up to 336 lb N/acre per year, ammonium sulphate also produced less palatable herbage than did calcium nitrate in the investigation reported by Widdowson *et al.* (494). Five forms of fertilizer N, each at 100 lb N/acre, were compared in the two trials carried out by Reid *et al.* (394) and described above. Again, there were differences between the spring hay and autumn grazing trials. In the spring hay trial, palatability decreased in the sequence: sodium nitrate > ammonium nitrate > ammonium sulphate > urea = ammonium phosphate. In the autumn-grazed trial, palatability decreased in the order: ammonium nitrate = ammonium phosphate > sodium nitrate > ammonium sulphate > urea. It has been suggested that the low palatability of grass receiving ammonium sulphate is due to its high S content (102).

Intake

Differences in palatability may not necessarily be reflected in differences in intake when only one type of herbage is available. In this situation, the level of fertilizer N has been found to have no effect on intake in a number of experiments (84, 85, 231, 319, 392, 394). However, Sonneveld (430) noted that sheep grazing pasture receiving 500 kg N/ha per year consumed less herbage than those grazing pasture fertilized with 100 kg N or given no N; and Bryant and Ulyatt (71) found in one of their two trials that sheep consumed less of the grass given a high level of N. On the other hand, Odhuba *et al.* (368) noted that increasing levels of fertilizer N significantly increased the intake of tall fescue by sheep. As pointed out by Raymond and Spedding (388), fertilizer N may increase intake in intensive grazing situations by making more herbage available. On the other hand, the resulting increase in moisture content (including superficial moisture) may sometimes reduce intake.

Digestibility

Generally, herbage digestibility is affected only slightly by the application of fertilizer N, at least with all-grass swards (132, 333, 388, 394). In some experiments, small increases (231, 392) or small decreases (29, 308) have been reported. However, the application of fertilizer N is likely to have a significant effect on herbage digestibility (a) where it delays herbage maturation, (b) where it reduces the proportion of clover in mixed swards, and (c) where, in its absence, soil N supplies are so low that herbage digestibility is reduced by very low herbage N contents (388).

Metabolic utilization

Very high N contents in herbage, particularly when associated with large proportions of non-protein N, may sometimes have unfavourable effects.

It has been suggested that high N contents in herbage may result in levels of ammonia in the rumen high enough to reduce rumen motility, with consequent decrease in appetite (166).

The presence of appreciable quantities of nitrate may also affect microbial fermentation in the rumen. In artificial-rumen studies, Perez and Story (379) showed that as nitrate-N content increased up to 0·72% in cocksfoot receiving 800 lb N/acre, fermentation, as assessed by total gas production, decreased to 40% of that from unfertilized herbage.

A number of workers have put forward the view that if herbage N is to be utilized efficiently, the available carbohydrate: N balance in the herbage must be such as to meet both the N and the energy requirements of the rumen micro-organisms (71, 226, 387). Heavy applications of N may upset this balance by increasing the total N content of herbage and at the same time decreasing its content of soluble carbohydrate.

The balance between N and S in ruminant diets may also be important, since effective utilization of the N involves its conversion to microbial protein, which has a relatively constant N : S ratio. Moir (335) has obtained evidence that the N : S ratio of a ruminant diet should not be wider than 10:1.

The optimum dietary N content on a DM basis for most types of ruminant is probably in the range 2·2–2·8% (198, 387).

CHAPTER 34

EFFECTS ON ANIMAL PRODUCTION

At present, the most reliable means of assessing the nutritive value of swards receiving various rates of fertilizer N is to carry out long-term feeding trials with livestock. When grazed herbage is the sole diet, it is particularly important that it should contain sufficient nutrients and in the right proportions. However, when livestock are fed freshly cut or conserved herbage, it is easier to supplement their diet with extra energy, N or minerals, as required; hence the need for it to be nutritionally well balanced is reduced.

The responses to fertilizer N obtained in animal production experiments have been reviewed recently by Holmes (234). He concluded that for dairy cows, milk production per hectare increased almost linearly over the range 0–450 kg N/ha per year, each kg N yielding about 1·05 cow-grazing days and about 15 kg milk. The data from experiments involving beef animals, however, indicated that while applications of up to about 200 kg N/ha gave responses of about 1 kg liveweight gain per kg N, the response was less at higher levels of N. Holmes pointed out that since animal production is influenced by many management factors other than rate of fertilizer N, it is not possible at present to state whether these results represent a genuine difference between the two classes of stock.

In general, one would expect animal production per unit area to be closely correlated with yield of herbage DM per unit area assuming herbage of constant quality. Direct proportional relationships between herbage yield and grazing days/ha and between grazing days/ha and milk production or liveweight gain/ha were postulated by Holmes (234) (see Table 15) in his proposed targets for animal production from grassland receiving various levels of fertilizer N. These targets for animal production do, however, assume rather higher herbage yields than those obtained in many experiments. The yield of 5500 kg DM/

TABLE 15

Targets for animal production from grass (**Holmes** (**234**))

Level of production	Fertilizer, kg N/ha	Herbage yields DM, kg/ha	Herbage yields DCP, kg/ha*	Cow-grazing† days/ha	Beef-grazing days/ha‡	Expected production in 180-day grazing season			
						Cows/ha	Milk kg/ha	Beef cattle /ha	Liveweight gain kg/ha
Low	0	5500	600	495 (392)	820	2·8 (2·2)	7430 (5880)	4·6	820
Medium	150	9800	1100	885 (700)	1460	4·9 (3·9)	13300 (10500)	8·1	1460
High	300	11700	1600	1050 (835)	1750	5·8 (4·6)	15700 (12500)	9·7	1750
Very high	450	13200	2020	1190 (940)	1970	6·6 (5·2)	17850 (14100)	10·9	1970

* Digestible crude protein, assuming 75% digestibility of crude protein
† 500-kg cow yielding 15 kg milk. 11·1 kg DM/day allowing for no gain in weight or, in brackets, 14 kg DM/day from direct intake measurements
‡ 350-kg cattle gaining 1 kg/day, 6·7 kg DM/day

ha with no fertilizer N presumably refers to a grass/clover sward; if so, the yield of 9800 kg DM/ha with an application of 150 kg N/ha is equivalent to a yield response of 28 kg DM/kg N, a much higher response than is usually obtained from grass/clover swards.

Although yield increases resulting from the application of high levels of fertilizer N should always be reflected in increased animal production per unit area, there is some evidence that the individual animals perform less well on grass swards heavily fertilized with N than on grass/clover swards given no N. Blaser (50) has quoted several N. American investigations showing that output per animal in terms of milk production or liveweight gain from cattle grazing grass/clover mixtures not given fertilizer N was usually higher than from cattle grazing N-fertilized grass. In a trial with sheep in New Zealand in which herbage supply was not a limiting factor, in 3 out of 4 comparisons grass/clover swards produced animals of 8–15% greater liveweight after 6 months than did all-grass swards receiving ammonium sulphate (383). White clover and lucerne not receiving fertilizer N also produced considerably greater liveweight gains in lambs than did 3 grass species receiving 200-300 lb N/acre, when plots of each species were grazed individually (314). Recent work by Grimes *et al.* (202) has also shown greater liveweight gains in lambs grazing grass/clover plots than in those grazing all-grass plots. Browne (70), however, found no significant differences in the liveweight gains of individual bullocks grazing mixed swards receiving different levels of N in the range 0–368 lb/acre. Also, Tayler and Rudman (443) found that the rate of liveweight gain in beef animals consuming grass/clover herbage was not reduced by the application of 208 lb N/acre per year, and that no differences in carcass conformation resulted from the fertilizer application.

With dairy cows, Campling *et al.* (86) found that an annual application of 212 lb N/acre to grass had no appreciable effect on milk yield per cow compared with a grass/clover sward which received no fertilizer N. There was no significant effect on the butterfat content of the milk, but solids-not-fat content was slightly reduced when fertilizer N was applied. In later work, Arnold and Holmes (19) found that fertilizer N at a rate of 202 lb/acre per year had no significant effect on milk yield or on contents of total solids, solids-not-fat, or butter-fat.

Sheep and cattle appear to differ in their responses to all-grass and grass/clover swards. Sheep are resistant to bloat and are consequently able to thrive on herbage rich in clover. In contrast, Alder *et al.* (6) found that bullocks had relatively low herbage intakes and made correspondingly low weight gains when grazing swards with very high contents of white clover (60–80% in herbage samples) and which tended to produce bloat.

It is possible that some of the differences in animal performance that have been reported have been due to inherent differences in composition between grasses and clovers, the presence of clover, at least with lambs, giving an advantage in terms of animal production so long as bloat is avoided.

With all-grass swards, high levels of fertilizer N appear to have had no deleterious effects on animal performance. Blaser (50) cited several investigations carried out in the USA showing no deleterious effects of various levels of fertilizer N up to 200–400 lb/acre per year on liveweight gain or milk production from all-grass swards. Bryant and Ulyatt (71) reported that sheep fed (for only 16 days) on grass from plots receiving 308 lb nitro-lime/acre per month showed no unfavourable effects on health or feed intake compared with sheep fed grass given 140 lb nitro-lime per month. Browne (67) reported that herbage from pastures receiving 369 lb N/acre per year had no detrimental effects on the performance or health of bullocks compared with that given lower levels of fertilizer N. Also, Hodgson and Spedding (228) and Large and Spedding (282) found that grass receiving 900–1000 lb N/acre per year as Nitro-chalk caused no deterioration in the productivity or health of calves or lambs compared with grass receiving 180–200 lb N.

CHAPTER 35

FERTILIZER NITROGEN AND METABOLIC DISORDERS OF LIVESTOCK

Most experiments which have investigated the effects of fertilizer N on animal performance have shown few deleterious effects where high rates have been used, and experience at the Nitrogen Experimental Farms in the Netherlands, on which annual N rates were 220–500 lb/acre, indicated that, in general, animal health was as good as on the average Dutch farm (203). However, metabolic disorders have sometimes apparently been associated with the use of high rates of fertilizer N, and there is still some uncertainty about the effects of the continued use of high rates over a period of years. It is possible that the application of fertilizer N might bring about a deficient or toxic level of some herbage constituent, and Large (281) has suggested that the reduction in herbage DM content caused by fertilizer N may trigger-off metabolic disturbances.

Evidence for the implication of fertilizer N in various disorders of animals is summarized below.

Nitrate toxicity

Although the nitrate ion itself is relatively non-toxic to animals, it is converted in the rumen to nitrite, which is then subject to further reduction to ammonia. In some circumstances the reduction of nitrite to ammonia takes place more slowly than normal (30), and nitrite, which is toxic, is then absorbed into the bloodstream. There is evidence (30, 168, 426) that nitrite accumulates in the rumen to a greater extent when animals consume diets low in available-carbohydrate content.

High levels of nitrate in fodders other than grass have on occasion resulted in the death of livestock. Wright and Davison (514), who reviewed the question of nitrate accumulation in crops and nitrate poisoning in animals, concluded that a level of about 0·4% nitrate N in forages should be regarded as potentially toxic to ruminants. However, Bryant and Ulyatt (71) fed grass with contents up to 0·72% nitrate N to sheep and found no apparent deleterious effects on health or feed intake over periods of 16 days. Also, Large and Spedding (282) observed no signs of poisoning in lambs grazing herbage containing 0·3–0·6% nitrate N throughout most of the season.

It has been suggested that levels of nitrate too low to cause acute toxicity might nevertheless result in metabolic disorders, but evidence for this is inconclusive (514). Evidence that nitrate can result in a depletion of vitamin A and that it may possibly interfere with iodine metabolism has been summarized by Garner (168) and Wright and Davison (514).

Ill-thrift in sheep

This condition involves loss of bodyweight and is common in New Zealand, particularly in hoggets (sheep in their second year) during autumn. In the moist, warm conditions at that time, grass grows rapidly and has been shown to contain relatively high amounts of non-protein N, particularly nitrate, coupled with relatively low levels of soluble carbohydrate, even when no fertilizer N has been applied. Although there is considerable evidence for a positive correlation between herbage nitrate content and symptoms of ill-thrift, workers in New Zealand (353) consider that nitrate is only one factor implicated in the condition. Other non-protein N constituents of herbage are being studied, as is the possibility that the microbial digestion of herbage with high N contents may produce toxic compounds in the rumen.

Hypomagnesaemia

The relationship between fertilizer N and hypomagnesaemia is not fully understood, although certain points are clear.

It is well established that, in spring at least, clovers have a higher Mg content than associated grasses, and that fertilizer N reduces the proportion of clover in a mixed sward. N, however, especially in nitrate form, increases the Mg content of grass herbage, and such increases have often been reported for grass/clover swards.

Despite this increase in the total Mg content of herbage, there is considerable evidence that N fertilizers have increased the incidence of hypomagnesaemia in some situations, particularly when K has also been applied (218, 266, 328, 338, 474). However, some workers have found no effect of fertilizer treatment on Mg levels in the blood of animals consuming the herbage. Thus Hemingway *et al.* (217) reported that 118 lb N/acre as Nitro-chalk, with or without 112 lb K, had no effect on plasma Mg levels in sheep over a 2-month period. Hodgson and Spedding (228) and Large and Spedding (282) found no differences in serum Mg levels between calves and lambs grazing grass receiving 800–1000 lb N/acre per year, compared with those on grass receiving 180–200 lb N. In these experiments, a basic dressing of 67 lb K/acre was applied during the winter preceding the first year of the trials. Also, L'Estrange *et al.* (288) reported no differences in serum Mg levels in ewes fed on herbage from mixed swards given either 100 lb N/acre in March and 60 lb at monthly intervals thereafter, or 40 lb N in March and none thereafter. On the other hand, Hvidsten (245) found that N applied at 185 kg/ha accentuated the depression in serum Mg levels in sheep caused by K applied at 164 kg/ha.

The evidence on the effect of fertilizer N on the availability of herbage Mg to animals, as assessed by balance trials, is also conflicting. L'Estrange *et al.* (289) found no difference in the apparent availability to sheep of the Mg in herbage receiving the two fertilizer treatments described above, whereas Stillings *et al.* (440) found that the apparent availability to sheep of the Mg in cocksfoot herbage receiving 50 lb fertilizer N/acre was 18–24%, but only 11–16% in herbage receiving 500 lb N/acre. The herbages contained 2·1–2·6% and 3·7–4·4% N, respectively.

Studies of the composition of rumen contents and the feeding of various supplements have also yielded inconclusive results. Head and Rook (212) reported that the serum Mg levels in dairy cows decreased as the levels of ammonia in the rumen increased at the beginning of the spring grazing period. They also found a decrease in serum Mg level following the introduction of ammonium salts into the rumen. Wilson (503), however, failed to induce a fall in the serum Mg level of sheep by feeding urea, despite a marked increase in the ammonia content of the rumen liquor, indicating that ammonia in itself did not affect Mg availability.

The consumption of herbage with a high N content would, of course, liberate N compounds other than ammonia into the rumen, and these may influence serum Mg levels. Larvor and Guéguen (284) reported a significant correlation between the soluble-N content of herbage and levels of serum Mg in grazing dairy cows. Ashton and Sinclair (21) suggested that compounds such as amino acids and amides, which have chelating properties, might be responsible for the decreased availability of Mg. The stability of such Mg chelates would be greater at the high pH values induced by large amounts of ammonia. Some support for this hypothesis was provided by the finding that when both EDTA and ammonium carbonate were fed to sheep, there was a decrease in the level of serum Mg (21).

An alternative mechanism, suggested by Kemp *et al.* (267), is that fertilizer N may reduce Mg availability by increasing the fat content of herbage. A linear correlation was found between the N content of herbage and the content of higher fatty acids and, in a feeding experiment with dairy cows, it was found that adding fat to the ration tended to reduce the retention of dietary Mg. It was suggested that excretion of Mg in the faeces was increased by the formation of insoluble Mg soaps.

CONCLUSION

The effective utilization of both natural and industrial sources of N requires a full understanding of the transformations which N undergoes in the soil, plant and animal components of grassland systems.

The importance of many of the transformations depends to a large extent on the type of grassland management. The sequence of transformations comes nearest to being a closed cycle under conditions where grass/clover swards receiving no inputs of fertilizer N are grazed extensively. Under such conditions, there is an approximate balance between the various losses of N from the system (by the removal of livestock products, volatilization of ammonia from animal excreta, denitrification, chemical reactions involving loss of gaseous N, and leaching) and additions from the atmosphere (by symbiotic and non-symbiotic fixation and in precipitation).

In contrast, in highly intensive production systems in which all-grass herbage is cut and removed from the sward, the main transformation is the uptake of fertilizer N by the grass, other transformations being significant mainly in so far as they result in a loss of fertilizer N in gaseous form or through leaching. In general, the transformations resulting in losses from the sward become increasingly important as the rate of N input is increased. Further investigation is required into the factors governing denitrification and similar chemical reactions resulting in losses of gaseous N from heavily fertilized swards.

In systems intermediate between extensive grazing and intensive grass production for cutting, the relationships between the various transformations are more complex. The grass component may receive N from clover, fertilizer, soil and/or from animal excreta. Changes in the supply from one source of N may influence the contribution made by one or more of the other sources, and the extent to which fertilizer N, clover N and N from excreta can be used in conjunction with one another is not well understood and requires further investigation. N applied as liquid manure or slurry appears to cause less depression of clover than fertilizer N and, at the same time, provides a readily available source of N for grass growth. The N returned in the excreta of grazing animals, although variable in its effect, appears to be generally much less effective than that in slurry. Losses of N by leaching, by the volatilization of ammonia and through denitrification are greater in grazing situations owing to the uneven distribution of the excreta over the sward.

Whether the major source of N should be clover or fertilizer, or some combination of both these sources, is the subject of continued discussion. Important considerations are : (1) the suitability of climate and soil for clover growth, (2) the output required per unit area, (3) the cost of fertilizer N in relation to the value to the farmer of the expected increase in herbage yield, (4) the greater predictability of herbage yield from grass receiving fertilizer N compared with grass/clover receiving no fertilizer N, and (5) the ability of fertilizer N to induce early growth in spring.

In New Zealand, where conditions for clover growth are favourable, clover is by far the most important source of N and seems likely to remain so in the foreseeable future (156). In the Netherlands, on the other hand, where conditions for clover growth are much less favourable and where much emphasis is placed on high output per unit area and fertilizer N is relatively cheap, fertilizer is the main source of N (137, 372).

In Britain, there is a considerable difference between the lowland areas where fertilizer N is becoming increasingly important, and the generally wetter upland areas where clover is relatively more important. For any one farm, the choice will be much influenced by local economic factors and by the relative emphasis placed on grazing and grass conservation. Although the combined use of fertilizer N and clover N is sometimes recommended, this inevitably involves difficulties. For example, the yield response of grass/clover swards to rates of N up to about 150 lb/acre per year is unreliable, and in some years may be negligible; thus if higher rates are to be used, the advantage to be gained by including clover in the sward may be offset by the increased problems of sward management. Reliance on clover during the early years of a sward's life, then changing to high rates of fertilizer N as the vigour of the clover declines, is also difficult to implement without incurring a year of relatively low production.

The response of grass to fertilizer N in terms of increased yield per unit of N applied is, of course, modified by various environmental factors. Thus, while there is evidence that fertilizer N can stimulate grass growth at temperatures between 5 and 9°C, yield responses to N are low, and below 5°C are negligible. Similarly, responses to N in temperate regions appear to be restricted when incoming solar radiation is less than about 550 cal/cm^2 per day.

These factors partly account for differences in response to fertilizer N applied at different times of the growing season. In early spring, responses are generally limited by low temperatures, but in autumn low light intensity appears to be a major limitation. However, the physiological condition of the plant is also important. Growth rates and responses to N are generally greatest in late spring when infloresences are being formed, and are often relatively low in midsummer, even when growth conditions appear to be favourable.

The seasonal pattern of growth and N response can, however, be modified by the frequency of defoliation. Maximum annual yields and responses are obtained in most temperate regions with infrequent defoliation (2–3 times per year), but in order to obtain more even distribution of production and better herbage quality, more frequent defoliation (4–8 times per year) is normal. Although much of the available N is taken up within 2 or 3 weeks of application under conditions which favour grass growth, a considerably longer period is required for the uptake to be fully reflected in DM production. N absorption by grass is rapid and there is normally very little carry-over of readily available N from one growth period to the next. Fertilizer N is therefore usually applied after each harvest for the succeeding growth period. More information is required on the factors governing the response of grass to N applied in mid- and late summer since the relative importance of environmental factors and of stage of growth is still unresolved.

Deficiencies of water and nutrients will, of course, restrict grass growth and response to N; and levels of supply which are sufficient in the absence of fertilizer N may be inadequate for the increased rate of growth resulting from the application of fertilizer N. For any one application of N, the timing of rainfall (or irrigation) is of considerable importance, especially when the topsoil is dry.

The solid N fertilizers in common use generally produce similar yield responses, though low responses to urea have been reported in a few trials and responses to ammonium salts may be low on calcareous soils owing to the loss of ammonia. In comparison with solid fertilizers, responses to injected anhydrous and aqueous ammonia have varied considerably between different trials, and further information is required on the ability of these forms of fertilizer N to provide a continuous supply of N over the growing season.

Although fertilizer N may cause substantial changes in the chemical composition of herbage, there is very little evidence that even high rates of application adversely affect livestock health. With grass/clover swards, however, changes in chemical composition inevitably result from the reduction in the clover content of the herbage.

With all-grass swards, high rates of fertilizer N greatly increase the contents of total N and, particularly for periods of up to about 3 weeks after application, the contents of nitrate N also. The content of soluble carbohydrate is often considerably decreased but there is little effect on cellulose or lignin. The effect of fertilizer N on contents of mineral elements in herbage is very variable. Some elements, including P and K, appear to be increased by fertilizer N when their supply in the soil is good but are decreased when in short supply. Herbage Mg is often slightly increased by fertilizer N, but there is some evidence that its availability to animals may be reduced. Both increases and decreases in herbage Cu content, apparently unrelated to Cu supply, have been reported. The Ca in Nitro-chalk and the S in ammonium sulphate tend to increase the contents of these elements in the herbage. Some forms of fertilizer N increase soil acidity whereas others do not, and this difference is reflected in their effects on herbage contents of elements, such as Mn, whose uptake is much influenced by soil pH.

The use of fertilizer N on grazed swards can increase stocking rates to an extent approximately proportional to herbage yield if stock management enables the increased yield to be utilized effectively. Although there is some evidence that lamb growth rates are higher on grass/clover than on grass fertilized with N, there appears to be no direct evidence that fertilizer N affects the performance of individual animals on all-grass swards. It is possible, however, that fertilizer N may, in some circumstances, increase the risk of certain metabolic disorders such as hypomagnesaemia and ill-thrift.

APPENDIX

Botanical names of species mentioned in the text

Bent	*Agrostis* sp.
Bermudagrass	*Cynodon dactylon*
Bluegrass	*Poa pratensis*
Bromegrass	*Bromus* sp.
Coastal Bermudagrass	*Cynodon dactylon*
Cocksfoot	*Dactylis glomerata*
Italian ryegrass	*Lolium multiflorum*
Ladino clover	*Trifolium repens*
Lucerne	*Medicago sativa*
Meadow fescue	*Festuca pratensis*
Perennial ryegrass	*Lolium perenne*
Red clover	*Trifolium pratense*
Red fescue	*Festuca rubra*
Reed canarygrass	*Phalaris arundinacea*
Rhodes grass	*Chloris gayana*
Subterranean clover	*Trifolium subterraneum*
Sudan grass	*Sorghum sudanense*
Tall fescue	*Festuca arundinacea*
Timothy	*Phleum pratense*
White clover	*Trifolium repens*
Wimmera ryegrass	*Lolium rigidum*
Yorkshire fog	*Holcus lanatus*

BIBLIOGRAPHY

1 ADAMS, C. A. ; SHEARD, R. W. Alterations in the nitrogen metabolism of *Medicago sativa* and *Dactylis glomerata* as influenced by potassium and sulfur nutrition. *Can. J. Pl. Sci.* 1966, **46,** 671–80.

2 ADAMS, J. R.; ANDERSON, M. S.; HULBERT, W. C. Liquid nitrogen fertilizers for direct application. *Agric. Hdbk 198 USDA* 1965, pp. 46.

3 ADAMS, W. E.; TWERSKY, M. Effect of soil fertility on winter killing of Coastal Bermudagrass. *Agron. J.* 1960, **52,** 325–6.

4 AGRICULTURAL RESEARCH COUNCIL. The nutrient requirements of farm livestock. No. 2. Ruminants. London : H.M.S.O., 1965.

5 ALBERDA, T. Some aspects of nitrogen in plants, more specially in grass. *Stikstof* (English edn) 1968, No. 12, 97–103.

6 ALDER, F. E.; COWLISHAW, S. J.; NEWTON, J. E.; CHAMBERS, D. T. The effects of level of nitrogen fertilizer on beef production from grazed perennial ryegrass/white clover pastures. Part 1. *J. Br. Grassld Soc.* 1967, **22**, 194–203.

7 ALDERMAN, G.; JONES, D. I. H. The iodine content of pastures. *J. Sci. Fd Agric.* 1967, **18,** 197–9.

8 ALEXANDER, M. Nitrification. In: Soil Nitrogen. Agronomy, No. 10. Bartholomew, W. V.; Clark, F. E. [Eds] Madison, Wisconsin: Amer. Soc. Agron., 1965, pp. 307–43.

9 ALLEN, S. E. The development of new nitrogen fertilizers as determined by the nutrient requirements of agronomic crops. *Proc. Nitrogen Res. Symp. Tennesse Valley Authority Div. Agric. Dev.* 1964.

10 ALLISON, F. E. The enigma of soil nitrogen balance sheets. *Adv. Agron.* 1955, **7,** 213–50.

11 ALLISON, F. E. Losses of gaseous nitrogen from soil by chemical mechanisms involving nitrous acid and nitrites. *Soil Sci.* 1963, **96,** 404–9.

12 ALLISON, F. E. Evaluation of incoming and outgoing processes that affect soil nitrogen. In: Soil Nitrogen. Agronomy, No. 10. Bartholomew, W. V.; Clark, F. E. [Eds] Madison, Wisconsin: Amer. Soc. Agron., 1965, pp. 573–606.

13 ALLISON, F. E. The fate of nitrogen applied to soils. *Adv. Agron.* 1966, **18,** 219–58.

14 ALLISON, F. E.; KEFAUVER, M.; ROLLER, E. M. Ammonium fixation in soils. *Proc. Soil Sci. Soc. Am.* 1953, **17,** 107–10.

15 ALLOS, H. F.; BARTHOLOMEW, W. V. Replacement of symbiotic fixation by available nitrogen. *Soil Sci.* 1959, **87,** 61–6.

16 ANSLOW, R. C.; GREEN, J. O. The seasonal growth of pasture grasses. *J. agric. Sci., Camb.* 1967, **68,** 109–22.

17 ANTONOVICS, J.; LOVETT, J.; BRADSHAW, A. D. The evolution of adaptation to nutritional factors in populations of herbage plants. In : International Atomic Energy Agency. Isotopes in plant nutrition and physiology. Proceedings of a Symposium, Vienna, 1966, IAEA and FAO. Vienna : IAEA, 1967, pp. 549–67.

18 ARMITAGE, E. R.; TEMPLEMAN, W. G. Response of grassland to nitrogenous fertilizer in the west of England. *J. Br. Grassld Soc.* 1964, **19,** 291–7.

19 ARNOLD, G. W.; HOLMES, W. Studies in grazing management. 7. The influence of strip grazing versus controlled free grazing in milk yield, milk composition and pasture utilization. *J. agric. Sci., Camb.* 1958, **51,** 248–56.

20 ARNOLD, P. W. Losses of nitrous oxide from soil. *J. Soil Sci.* 1954, **5,** 116–28.

21 ASHTON, W. M.; SINCLAIR, K. B. A study of the possible role of chelation in the occurrence of hypomagnesaemia in sheep. *J. Br. Grassld Soc.* 1965, **20,** 118–22.

22 ATKINSON, W. T.; WALKER, M. H.; WEIR, R. G. The phosphorus and sulphur needs of pasture in New South Wales. *Proc. 9th int. Grassld Congr., São Paulo, 1965* 1966, pp. 655–63.

23 AUDA, H.; BLASER, R. E.; BROWN, R. H. Tillering and carbohydrate contents of orchardgrass as influenced by environmental factors. *Crop Sci.* 1966, **6,** 139–43.

24 BAKER, H. K.; DAVID, G. L. Winter damage to grass. *Agriculture, Lond.* 1963, **70,** 380–2.

25 BAKER, H. K.; GARWOOD, E. A. Studies in the root development of herbage plants. 4. Seasonal changes in the root and stubble weights of various leys. *J. Br. Grassld Soc.* 1959, **14,** 94–104.

26 BAKER, H. K.; JONES, L.; CHARD, J. R. A. The control of weeds by MCPA in permanent pasture under different managements and the effects on herbage productivity. *Proc. 5th Br. Weed Control Conf.* 1960, **1,** 141.

27 BAKHUIS, J. A.; KLETER, H. J. Some effects of associated growth on grass and clover under field conditions. *Neth. J. agric. Sci.* 1965, **13,** 280–310.

28 BARLEY, K. P. Earthworms and the decay of plant litter and dung—a review. *Proc. Aust. Soc. Anim. Prod.* 1964, **5,** 236–40.

29 BARLOW, C. The use of special-purpose grasses for silage production. 1. A comparison of varieties. *J. agric. Sci., Camb.* 1965, **64,** 439–47.

30 BARNET, A. J. G.; BOWMAN, I. B. R. *In vitro* studies on the reduction of nitrate by rumen liquor. *J. Sci. Fd Agric.* 1957, **8,** 243–8.

31 BARROW, N. J. Some aspects of the effects of grazing on the nutrition of pastures. *J. Aust. Inst. agric. Sci.* 1967, **33,** 254–62.

32 BARROW, N. J. ; JENKINSON, D. S. The effect of water-logging on fixation of nitrogen by soil incubated with straw. *Pl. Soil* 1962, **16,** 258-62.

33 BARROW, N. J.; LAMBOURNE, L. J. Partition of excreted nitrogen, sulphur and phosphorus between the faeces and urine of sheep being fed pasture. *Aust. J. agric. Res.* 1962, **13,** 461–71.

34 BATHURST, N. O.; MITCHELL, K. J. The effect of light and temperature on the chemical composition of pasture plants. *N.Z. Jl agric. Res.* 1958, **1,** 540–52.

35 BEAN, E. W. The influence of nitrogen and light intensity on the growth of grass swards. Ph.D. thesis, Univ. Reading, 1961. (Seen in *Herb. Abstr.* **32** : 353)

36 BEATON, J. D.; HUBBARD, W. A.; SPIER, R. C. Coated urea, thiourea, urea-formaldehyde, hexamine, oxamide, glycol-uril and oxidised nitrogen-enriched coal as slowly available sources of nitrogen for orchard-grass. *Agron. J.* 1967, **59,** 127–33.

37 BENNET, W. D. A note on the effect of nitrate and phosphate on the perloline content of perennial ryegrass. *N.Z. Jl agric. Res.* 1963, **6,** 310–13.

38 BERLIER, Y. ; GUIRAUD, G. [Uptake and use by Gramineae of nitrogen from ^{15}N-labelled nitrate or ammonia.] In : International Atomic Energy Agency. Isotopes in plant nutrition and physiology. Proceedings of a Symposium, Vienna, 1966, IAEA and FAO. Vienna : IAEA, 1967, pp. 145–57.

39 BERRYMAN, C. Composition of organic manures and waste products used in agriculture. *Advis. Pap. 2 N.A.A.S., Minist. Agric., Fish., Fd* 1965.

40 BEZEAU, L. M.; LUTWICK, L. E.; SMITH, A. D.; JOHNSTON, A. Effect of fertilization on chemical composition, nutritive value and silica content of rough fescue, *Festuca scabrella. Can. J. Pl. Sci.* 1967, **47,** 269–72.

41 BIZZELL, J. A. Lysimeter experiments. 6. The effects of cropping and fertilization on the losses of nitrogen from the soil. *Mem.* 256 *N.Y. St. Coll. Agric. Cornell* 1944, pp. 14.

42 BLACK, W. J. M.; RICHARDS, R. I. W. A. Grassland fertilizer practice and hypomagnesaemia. *J. Br. Grassld Soc.* 1965, **20,** 110–17.

43 BLACKBURN, T. H. Nitrogen metabolism in the rumen. In : Physiology of digestion in the ruminant. Dougherty, R. W. *et al.* [Eds] London : Butterworth and Co. Ltd, 1965, pp. 322-34.

44 BLACKMAN, G. E. The influence of temperature and available nitrogen supply on the growth of pasture in spring. *J. agric. Sci., Camb.* 1936, **36,** 620–47.

45 BLACKMAN, G. E.; TEMPLEMAN, W. G. The interaction of light intensity and nitrogen supply in the growth and metabolism of grasses and clover (*Trifolium repens*). 2. The influence of light intensity and nitrogen supply on the leaf production of frequently defoliated plants. *Ann. Bot.* 1938, **2,** 765–91.

46 BLACKMAN, G. E.; TEMPLEMAN, W. G. The interaction of light intensity and nitrogen supply in the growth and metabolism of grasses and clover (*Trifolium repens*). 4. The relation of light intensity and nitrogen supply to the protein metabolism of the leaves of grasses. *Ann. Bot.* 1940, **4,** 533–87.

47 BLAGDEN, P. The potassium cycle in grassland. Ph.D. thesis, Univ. Nottingham, 1969.

48 BLAKE, G. A. Liquid manure : the fertilizer value of fertilizer washings. *N.Z. Jl Agric.* 1942, **65,** 257–9.

49 BLAND, B. F. The effect of cutting frequency and root segregation on the yield from perennial ryegrass—white clover associations. *J. agric. Sci., Camb.* 1967, **69,** 391–7.

50 BLASER, R. E. Symposium on forage utilization : effects of fertility levels and stage of maturity on forage nutritive values. *J. Anim. Sci.* 1964, **23,** 246–53.

51 BLASER, R. E.; BRADY, N. C. Nutrient competition in plant associations. *Agron. J.* 1950, **42,** 128–35.

52 BOSCH, S. [Grassland and nitrogen manuring.] *Tijdschr. ned. Heidemaatsch.* 1964, **75,** 261–6. (Seen in *Soils Fertil., Harpenden* **28** : 1338).

53 BOYD, D. A. Estimating trends in fertilizer use. *Rep. Rothamsted exp. Stn 1966* 1967, pp. 339–48.

54 BREMNER, J. M. Nitrogen availability indexes. In : Methods of Soil Analysis. Agronomy, No. 9. Black, C. A. *et al.* [Eds] Madison, Wisconsin : Amer. Soc. Agron., 1965, pp. 1324–45.

55 BREMNER, J. M.; SHAW, K. Denitrification in soil. *J. agric. Sci., Camb.* 1958, **51,** 22–52.

56 BROADBENT, F. E. Biological and chemical aspects of mineralisation. *Trans. Joint Meet. Comm. IV and V Int. Soc. Soil Sci., New Zealand 1962* 1962, pp. 220–9.

57 BROADBENT, F. E. Effect of fertilizer nitrogen on the release of soil nitrogen. *Proc. Soil Sci. Soc. Am.* 1965, **29,** 692–6.

58 BROADBENT, F. E. Interchange between inorganic and organic nitrogen in soils. *Hilgardia* 1966, **37,** 165–80.

59 Broadbent, F. E.; Clark, F. Denitrification. In: Soil Nitrogen. Agronomy, No. 10. Bartholomew, W.V.; Clark, F. E. [Eds] Madison, Wisconsin : Amer. Soc. Agron., 1965, pp. 344–59.

60 Broadbent, F. E.; Hill, G. N.; Tyler, K. B. Transformations and movement of urea in soils. *Proc. Soil Sci. Soc. Am.* 1958, **22,** 303–7.

61 Broadbent, F. E.; Nakashima, T. Reversion of fertilizer nitrogen in soils. *Proc. Soil Sci. Soc. Am.* 1967, **31,** 648–52.

62 Brockman, J. S. The role of legumes in British grassland. *Chemy Ind.* 1962, No. 4, 765–7.

63 Brockman, J. S. The relationship between total N input and yield of cut grass. *J. Br. Grassld Soc.* 1969, **24,** 89–96.

64 Brockman, J. S.; Wolton, K. M. The use of nitrogen on grass/white clover swards. *J. Br. Grassld Soc.* 1963, **18,** 7–13.

65 Brouwer, R.; Jenneskens, P. J.; Borggreve, G. J. Growth responses of shoots and roots to interruptions of the nitrogen supply. *Jaarb. Inst. biol. scheik. Onderz. LandbGewass. 1961* 1961, pp. 29–36.

66 Brown, B. A.; Munsell, R. I. Clovers in permanent grassland as influenced by fertilization. *Bull. 329 Conn. agric. Exp. Stn* 1956.

67 Browne, D. The effect of nitrogen on beef production from reseeded and old permanent pastures. *Proc. 1st Gen. Meet. Eur. Grassld Fed.* 1965, pp. 183–92.

68 Browne, D. Residual effects of nitrogenous fertilizers on leys. *Res. Rep. agric. Inst. Dubl. (Soils Div.)* 1966, pp. 75–6.

69 Browne, D. Nitrogen use on grassland. 1. Effect of applied nitrogen on animal production from a ley. *Ir. J. agric. Res.* 1966, **5,** 89–101.

70 Browne, D. Nitrogen use on grassland. 2. Effect of applied nitrogen on animal production from an old permanent pasture. *Ir. J. agric. Res.* 1967, **6,** 73–81.

71 Bryant, A. M.; Ulyatt, M. J. Effects of nitrogenous fertilizer on the chemical composition of short-rotation ryegrass and its subsequent digestion by sheep. *N.Z. Jl agric. Res.* 1965, **8,** 109–17.

72 Bryant, H. T. ; Blaser, R. E. Yields and stands of orchardgrass compared under clipping and grazing intensities. *Agron. J.* 1961, **53,** 9–11.

73 Burg, P. F. J. van. Nitrogen fertilization and the seasonal production of grassland herbage. *Proc. 8th int. Grassld Congr., Reading, 1960* 1961, pp. 142–6.

74 Burg, P. F. J. van. Nitrogen fertilization of grassland. The effect of nitrogen solutions. *Stikstof* (English edn) 1964, No. 8, 28–32.

75 Burg, P. F. J. van. Nitrate as an indicator of the nitrogen-nutrition status of grass. *Proc. 10th int. Grassld Congr., Helsinki, 1966* 1966, pp. 267–72.

76 Burg, P. F. J. van. Nitrogen fertilizing of grassland in spring. *Neth. Nitrogen Tech. Bull.* 1968, No. 6, pp. 45.

77 Burg, P. F. J. van ; Brakel, G. D. van. The fertilizer value of anhydrous ammonia on permanent grassland. *Stikstof* (English edn) 1965, No. 9, 28–36.

78 Burg, P. F. J. van; Brakel, G. D. van; Schepers, J. H. The agricultural value of anhydrous ammonia on grassland: experiments 1963–1965. *Neth. Nitrogen Tech. Bull.* 1967, No. 2, pp. 31.

79 Burton, G. W.; Jackson, J. E. Effect of rate and frequency of applying six nitrogen sources on Coastal Bermudagrass. *Agron. J.* 1962, **54,** 40–3.

80 Burton, G. W.; Southwell, B. L.; Johnson, J. C. The palatability of Coastal Bermudagrass (*Cynodon dactylon* (L) Pers.) as influenced by nitrogen level and age. *Agron. J.* 1956, **48,** 360–2.

81 Butler, G. W. The chemical composition of rapidly growing ryegrass and its relation to animal production. *Proc. N.Z. Soc. Anim. Prod.* 1959, **19,** 99–110.

82 Butler, G. W.; Bathurst, N. O. The underground transference of nitrogen from clover to associated grass. *Proc. 7th int. Grassld Congr., New Zealand, 1956* 1957, pp. 168–78.

83 Butler, G. W.; Greenwood, R. M.; Soper, K. Effects of shading and defoliation on the turnover of root and nodule tissue of plants of *Trifolium repens, Trifolium pratense* and *Lotus uliginosus. N.Z. Jl agric. Res.* 1959, **2,** 415–26.

84 Cameron, C. D. T. The effects of nitrogen fertilizer application rates to grass on forage yields, body weight gains, feed utilization and vitamin A status of steers. *Can. J. Anim. Sci.* 1966, **46,** 19–24.

85 Cameron, C. D. T. Intake and digestibility of nitrogen-fertilized grass hays by wethers. *Can. J. Anim. Sci.* 1967, **47,** 123–5.

86 Campling, R. C.; MacLusky, D. S.; Holmes, W. Studies in grazing management. 6. The influence of free- and strip-grazing and of nitrogenous fertilizers on production from dairy cows. *J. agric. Sci., Camb.* 1958, **51,** 62–9.

87 Carey, V.; Mitchell, H. L.; Anderson, K. Effect of nitrogen fertilization on the chemical composition of bromegrass. *Agron. J.* 1952, **44,** 467–9.

88 Carr, A. J. H.; Catherall, P. L. The assessment of disease in herbage crops. *Rep. Welsh Pl. Breed. Stn, 1963* 1964, pp. 94–100.

89 Carroll, J. C. Effects of drought, temperature and nitrogen on turf grasses. *Pl. Physiol., Lancaster* 1943, **18,** 19–36.

90 Carter, L. P; Scholl, J. M. Effectiveness of inorganic nitrogen as a replacement for legumes grown in association with forage grasses. 1. Dry matter production and botanical composition. *Agron. J.* 1962, **54,** 161–3.

91 CASTLE, M. E. Some recent grassland experiments and their significance in British grassland farming. *Agric. Prog.* 1965, **40**, 35–41.

92 CASTLE, M. E.; DRYSDALE, A. D. Liquid manure as a grassland fertilizer. 1. The response to liquid manure and to dry fertilizer. *J. agric. Sci., Camb.* 1962, **58**, 165–71.

93 CASTLE, M. E.; HOLMES, W. The intensive production of herbage for crop drying. 7. The effect of further continued massive applications of nitrogen with and without phosphate and potash on the yield of grassland herbage. *J. agric. Sci., Camb.* 1960, **55**, 251–60.

94 CASTLE, M. E.; REID, D. Nitrogen and herbage production. *J. Br. Grassld Soc.* 1963, **18**, 1–6.

95 CASTLE, M. E.; REID, D.; HEDDLE, R. G. The effect of varying the date of application of fertilizer nitrogen on the yield and seasonal productivity of grassland. *J. agric. Sci., Camb.* 1965, **64**, 177–84.

96 CHABANNES, J.; BARBIER, G.; DRIARD, J. [More observations on the spring mineralization of soil nitrogen.] *C.r. hebd. Séanc. Acad. Agric. Fr.* 1964, **50**, 874–81. (Seen in *Soils Fertil., Harpenden* **28** : 699).

97 CHASE, F. E.; CORKE, C. T.; ROBINSON, J. B. Nitrifying bacteria in soil. In : The Ecology of Soil Bacteria. Gray, T. R. G.; Parkinson, D. [Eds] Liverpool : University Press, 1967, pp. 593–611.

98 CHESTNUTT, D. M. B. The effects of nitrogen, potash and phosphate on the yield and composition of a grass/clover sward. *Rec. agric. Res.* (*Nth. Ireld*) 1965, **14** (1), 71–81.

99 CLEMENT, C. R.; COWLING, D. W. Grassland Research Institute. Unpublished data.

100 CLEMENT, C. R.; WILLIAMS, T. E. Leys and soil organic matter. 2. The accumulation of nitrogen in soils under different leys. *J. agric. Sci., Camb.* 1967, **69**, 133–8.

101 COLLINS, D. P. Nitrogen application on grassland. Ph.D. thesis Univ. Melbourne, 1967.

102 CONROY, E. Effects of heavy applications of nitrogen on the composition of herbage. *Ir. J. agric. Res.* 1961, **1**, 67–71.

103 COOKE, G. W. Nitrogen fertilizers. Their place in food production, the forms which are made and their efficiencies. *Proc. Fertil. Soc.* 1964, No. 80, 1–88.

104 COOKE, G. W. The control of soil fertility. London : Crosby Lockwood & Son Ltd, 1967, pp. 526.

105 COOPER, C. S.; KLAGER, M. G.; SCHULZ-SCHAEFFER, J. Performance of six grass species under different irrigation and nitrogen treatments. *Agron. J.* 1962, **54**, 283–8.

106 COWLING, D. W. The effect of nitrogenous fertilizer on an established white clover sward. *J. Br. Grassld Soc.* 1961, **16**, 65–8.

107 Cowling, D. W. The effect of white clover and nitrogenous fertilizer on the production of a sward. 1. Total annual production. *J. Br. Grassld Soc.* 1961, **16,** 281–90.

108 Cowling, D. W. The effect of white clover and nitrogenous fertilizer on the production of a sward. 2. Seasonal production. *J. Br. Grassld Soc.* 1962, **17,** 282–6.

109 Cowling, D. W. Nitrogenous fertilizer and seasonal production. *J. Br. Grassld Soc.* 1963, **18,** 14–17.

110 Cowling, D. W. The response of grass swards to nitrogenous fertilizer. *Proc. 10th int. Grassld Congr., Helsinki, 1966* 1966, pp. 204–9.

111 Cowling, D. W. The effect of the early application of nitrogenous fertilizer and of the time of cutting in spring on the yield of rye-grass/white clover swards. *J. agric. Sci., Camb.* 1966, **66**, 413–31.

112 Cowling, D. W. Ammonia as a source of nitrogen for grass swards. *J. Br. Grassld Soc.* 1968, **23,** 53–60.

113 Cowling, D. W. Grassland Research Institute. Unpublished data.

114 Cowling, D. W.; Green, J. O.; Green, S. M. The effect of white clover and nitrogenous fertilizer on the production of a sward. 3. Statistical interpretation of their relative contributions. *J. Br. Grassld Soc.* 1964, **19,** 419–24.

115 Cowling, D. W.; Lockyer, D. R. A comparison of the reaction of different grass species to fertilizer nitrogen and to growth in association with white clover. 1. Yield of dry matter. *J. Br. Grassld Soc.* 1965, **20,** 197–204.

116 Cowling, D. W.; Lockyer, D. R. A comparison of the reaction of different grass species to fertilizer nitrogen and to growth in association with white clover. 2. Yield of nitrogen. *J. Br. Grassld Soc.* 1967, **22,** 53–61.

117 Crooks, P.; Spiers, R. B.; Salam, A. A. Anhydrous ammonia for grassland. *Exp. Wk Edinb. Sch. Agric. 1966* 1967, pp. 59–60.

118 Cummings, G. A.; Teel, M. R. Effect of nitrogen, potassium and plant age on certain nitrogenous constituents and malate content of orchardgrass (*Dactylis glomerata* L.). *Agron. J.* 1965, **57,** 127–9.

119 Cunningham, R. K.; Cooke, G. W. Soil nitrogen. 2. Changes in levels of inorganic nitrogen in a clay-loam soil caused by fertilizer additions, by leaching and uptake by grass. *J. Sci. Fd Agric.* 1958, **9,** 317–24.

120 Cunningham, R. K.; Nielsen, K. F. Cation-anion relationships in crop nutrition. 5. The effects of soil temperature, light intensity and soil-water tension. *J. agric. Sci., Camb.* 1965, **64,** 379–86.

121 Dale, W. R. Some effects of sheep urine on pasture. *Proc. N.Z. Grassld Ass. 1961* 1961, pp. 118–23.

122 D'Aoust, M. J. Some aspects of the interaction between nitrogen and water in the growth of grass swards. Ph.D. thesis, Univ. Reading, 1965.

123 D'Aoust, M. J.; Tayler, R. S. The interaction between nitrogen and water in the growth of grass swards. *J. agric. Sci., Camb.* 1968, **70,** 11–17.

124 Davidson, R. L. Theoretical aspects of nitrogen economy in grazing experiments. *J. Br. Grassld Soc.* 1964, **19,** 273–80.

125 Davies, D. D.; Giovanelli, J.; Rees, T. ap. Plant Biochemistry. Oxford : Blackwell Scientific Publications, 1964, pp. 454.

126 Davies, E. B.; Hogg, D. E.; Hopewell, H. G. Extent of return of nutrient elements by dairy cattle : possible leaching losses. *Trans. Joint Meet. Comm. IV and V Int. Soc. Soil Sci., New Zealand, 1962* 1962, pp. 715–20.

127 Davies, H.; Thomas, B. V.; Aldrich, D. T. A. The effects of different levels of nitrogen fertilizer on the yields of varieties of perennial ryegrass and cocksfoot. *J. Br. Grassld Soc.* 1966, **21,** 245–9.

128 Davies, W. E.; Davies, R. O.; Harvard, A. The yield and composition of lucerne, grass and clover under different systems of management. 3. The effect of nitrogen and frequency of cutting. *J. Br. Grassld Soc.* 1960, **15,** 106–15.

129 Davies, W. E.; Griffith, G. ap ; Ellington, A. The assessment of herbage legume varieties. 2. *In vitro* digestibility, water soluble carbohydrate, crude protein and mineral content of primary growth of clover and lucerne. *J. agric. Sci., Camb.* 1966, **66,** 351–7.

130 Deinum, B. Climate, nitrogen and grass. Research into the influence of light intensity, water supply and nitrogen on the production and chemical composition of grass. *Meded. LandbHogesch. Wageningen* 1966, **66** (11), 91.

131 Delwiche, C. C.; Wijler, J. Non-symbiotic nitrogen fixation in soil. *Pl. Soil* 1956, **7,** 113–29.

132 Dent, J. W.; Aldrich, D. T. A. Systematic testing of quality in grass varieties. 2. The effect of cutting dates, season and environment. *J. Br. Grassld Soc.* 1968, **23,** 13–19.

133 Devine, J. R.; Holmes, M. R. J. Field experiments comparing ammonium nitrate, ammonium sulphate and urea applied repetitively to grassland. *J. agric. Sci., Camb.* 1963, **60,** 297–304.

134 Devine, J. R.; Holmes, M. R. J. Field experiments comparing ammonium nitrate and ammonium sulphate as top-dressings for winter wheat and grassland. *J. agric. Sci., Camb.* 1964, **62,** 377–9.

135 DEVINE, J. R.; HOLMES, M. R. J. Field experiments comparing winter and spring applications of ammonium sulphate, ammonium nitrate, calcium nitrate, and urea to grassland. *J. agric. Sci., Camb.* 1965, **64,** 101–7.

136 DIJKSHOORN, W. Nitrogen, chlorine and potassium in perennial ryegrass and their relation to the mineral balance. *Neth. J. agric. Sci.* 1958, **6,** 131–8.

137 DILZ, K. The effect of nitrogen on a clover/grass mixture and its application in farming practices. *Stikstof* (English edn) 1965, No. 9, 36–45.

138 DILZ, K.; BURG, P. F. J. van. Nitrogen fertilization of grassland. Comparison of the nitrogen fertilizer materials urea and nitrolime. *Stikstof* (English edn) 1963, No. 7, 66–71.

139 DILZ, K.; MULDER, E. G. Effect of associated growth on yield and nitrogen content of legume and grass plants. *Pl. Soil* 1962, **16,** 229–37.

140 DILZ, K.; MULDER, E. G. The effect of soil pH, stable manure and fertilizer nitrogen on the growth of red clover and of red clover associations with perennial ryegrass. *Neth. J. agric. Sci.* 1962, **10,** 1–22.

141 DILZ, K.; WOLDENDORP, J. W. Distribution and nitrogen balance of ^{15}N-labelled nitrate applied on grass sods. *Proc. 8th int. Grassld Congr., Reading, 1960* 1961, pp. 150–2.

142 DOAK, B. W. Some chemical changes in the nitrogenous constituents of urine when voided on pasture. *J. agric. Sci., Camb.* 1952, **42,** 162–71.

143 DONALD, C. M. The impact of cheap nitrogen. *J. Aust. Inst. agric. Sci.* 1960, **26,** 319–38.

144 DONALD, C. M. Competition among crop and pasture plants. *Adv. Agron.* 1963, **15,** 1–118.

145 DONALD, C. M.; WILLIAMS, C. H. Fertility and productivity of a podzolic soil as influenced by subterranean clover (*Trifolium subterraneum* L.) and superphosphate. *Aust. J. agric. Res.* 1954, **5,** 664–87.

146 DÖRING, H. [Is it possible to increase the vitamin contents of plants by fertilizing ?] *Dt. Landwirt.* 1961, **12,** 69–71. (Seen in *Soils Fertil., Harpenden* **24** : 2034).

147 DOSS, B. D.; ASHLEY, D. A.; BENNETT, O. L.; PATTERSON, R. M. Interactions of soil moisture, nitrogen and clipping frequency on yield and nitrogen content of Coastal Bermudagrass. *Agron. J.* 1966, **58,** 510–12.

148 DOTZENKO, A. D.; HENDERSON, K. E. Performance of five orchardgrass varieties under different nitrogen treatments. *Agron. J.* 1964, **56,** 152–5.

149 DOUGHTY, J. L. The rate of decomposition of plant roots. *Scient. Agric.* 1941, **21,** 429–32.

150 DRAYCOTT, A. P.; HODGSON, D. R.; HOLLIDAY, R. Recent research on the value of fertilizers in solution. *Agric. Prog.* 1967, **42,** 68–81.

151 DRYSDALE, A. D. Liquid manure as a grassland fertilizer. 2. The response to winter applications. *J. agric. Sci., Camb.* 1963, **61,** 353–60.

152 DRYSDALE, A. D. Liquid manure as a grassland fertilizer. 3. The effect of liquid manure on the yield and botanical composition of pasture and its interaction with nitrogen, phosphate and potash fertilizers. *J. agric. Sci., Camb.* 1965, **65,** 333–40.

153 DRYSDALE, A. D. A comparison of two sources of nitrogen for grassland, with special reference to the grass/clover ratio. *Proc. 10th int. Grassld Congr., Helsinki, 1966* 1966, pp. 255–8.

154 DURING, C.; MCNAUGHT, K. J. Effects of cow urine on growth of pasture and uptake of nutrients. *N.Z. Jl agric. Res.* 1961, **4,** 591–605.

155 EKPETE, D. M.; CORNFIELD, A. H. Effect of pH and addition of organic materials on denitrification losses from soil. *Nature, Lond.* 1965, **208,** 1200.

156 ELLIOTT, I. L. Implications and perspectives from earlier assessments. *N.Z. agric. Sci.* 1966, **1,** No. 8, 15–18.

157 ENNIK, G. C. The influence of management and nitrogen application on the botanical composition of grassland. *Neth. J. agric. Sci.* 1965, **13,** 222–37.

158 ERIKSSON, E. Composition of atmospheric precipitation. 1. Nitrogen compounds. *Tellus* 1952, **4,** 215–32.

159 FAO. Production yearbook 1949. Rome: FAO, 1950, p. 175.

160 FAO. Production yearbook 1968. Rome : FAO, 1969, p. 458.

161 FERGUSON, W. S. Pasture chemistry. In: Animal health, production and pasture. Worden, A. N.; Sellars, K. C.; Tribe, D. E. [Eds] London: Longmans Green & Co. Ltd, 1963, pp. 92–127.

162 FERGUSON, W. S.; TERRY, R. A. The fractionation of the non-protein nitrogen of grassland herbage. *J. Sci. Fd Agric.* 1954, **5,** 515–24.

163 FRAME, J. The effects of cutting and grazing techniques on productivity of grass/clover swards. *Proc. 9th int. Grassld Congr., São Paulo, 1965* 1966, pp. 1511–6.

164 FRAME, J. Anhydrous ammonia fertilizer. *Scott. Agric.* 1967, **66,** 186–8.

165 FRENEY, J. R. An evaluation of naturally occurring fixed ammonium in soils. *J. agric. Sci., Camb.* 1964, **63,** 297–303.

166 FRENS, A. M. Physical and chemical aspects of the digestion of grass and grassland products by the ruminant. In : Chemical aspects of the production and use of grass. Soc. Chem. Ind. Monogr. No. 9. London : Soc. Chem. Ind., 1960, pp. 202–12.

167 Gardner, E. H.; Jackson, T. L.; Webster, G. R.; Turley, R. H. Some effects of fertilization on the yield, botanical and chemical composition of irrigated grass, and grass-clover pasture swards. *Can. J. Pl. Sci.* 1960, **40,** 546–62.

168 Garner, G. B. Nitrate—a factor in animal health. *Proc. N.Z. Soc. Anim. Prod.* 1963, **23,** 28–38.

169 Garwood, E. A. Some effects of soil water conditions and soil temperature on the roots of grasses. 1. The effect of irrigation on the weight of root material under various swards. *J. Br. Grassld Soc.* 1967, **22,** 176–81.

170 Garwood, E. A.; Clement, C. R.; Williams, T. E. Leys and soil organic matter. 3. The increase in macro-organic matter. (in preparation)

171 Garwood, E. A.; Williams, T. E. Growth, water use and nutrient uptake from the subsoil by grass swards. *J. agric. Sci., Camb.* 1967, **69,** 125–30.

172 Gasser, J. K. R. Soil nitrogen. 4. Transformations and movement of fertilizer nitrogen in a light soil. *J. Sci. Fd Agric.* 1959, **10,** 192–7.

173 Gasser, J. K. R. Some processes affecting nitrogen in the soil. In: Nitrogen and soil organic matter. *Tech. Bull. 15 Min. Agric., Fish., Fd* London: H.M.S.O., 1969, pp. 15–29.

174 Gasser, J. K. R. Urea as a fertilizer. *Soils Fertil., Harpenden* 1964, **27,** 175–80.

175 Gasser, J. K. R.; Greenland, D. J.; Rawson, R. A. G. Measurement of losses from fertilizer nitrogen during incubation in acid sandy soils and during subsequent growth of ryegrass using ^{15}N-labelled fertilizers. *J. Soil Sci.* 1967, **18,** 289–300.

176 Gasser, J. K. R.; Widdowson, F. V. Ammonia as a fertilizer. *J. Inst. Corn agric. Merch.* 1966, **14,** 130–3.

177 Gething, P. A. The effects of nitrogen, phosphate and potash on yields of herbage cut for conservation. *Proc. 1st Reg. Conf. int. Potash Inst., Wexford (Ireland) 1963* 1963, pp. 83–96.

178 Gillard, P. Coprophagous beetles in pasture ecosystems. *J. Aust. Inst. agric. Sci.* 1967, **33,** 30–4.

179 Gisiger, L. Organic manuring of grassland. *J. Br. Grassld Soc.* 1950, **5,** 63–79.

180 Goedewaagen, M. A. J.; Schuurman, J. J. [Root production in arable and grassland as a source of soil organic matter.] *Landbouwk. Tijdschr., Wageningen* 1950, **62,** 469–82.

181 Gordon, C. H.; Decker, A. M.; Wiseman, H. G. Some effects of nitrogen fertilizer, maturity and light on the composition of orchardgrass. *Agron. J.* 1962, **54,** 376–8.

182 Goss, R. L.; Gould, C. J. Some interrelationships between fertility levels and *Ophiobolus* patch disease on turfgrasses. *Agron. J.* 1967, **59,** 149–51.

183 Grassland Research Institute. *Exps Prog. Grassld Res. Inst. No. 14* 1962, p. 42.

184 Grassland Research Institute. *Exps Prog. Grassld Res. Inst. No. 15* 1963, pp. 17–18.

185 Grassland Research Institute. *Exps Prog. Grassld Res. Inst. No. 15* 1963, pp. 42–3.

186 Grassland Research Institute. *Exps Prog. Grassld Res. Inst. No. 16* 1964, pp. 22–3.

187 Grassland Research Institute. *Exps Prog. Grassld Res. Inst. No. 16* 1964, pp. 52–3.

188 Grassland Research Institute. *Exps Prog. Grassld Res. Inst. No. 17* 1965, pp. 22–3.

189 Gray, B.; Drake, M.; Colby, W. G. Potassium competition in grass-legume associations as a function of root cation exchange capacity. *Proc. Soil Sci. Soc. Am.* 1953, **17,** 235–9.

190 Great House Experimental Husbandry Farm. *Great House Review* 1966, pp. 24–6.

191 Great House Experimental Husbandry Farm. Unpublished data.

192 Green, J. O.; Corrall, A. J. Grassland Research Institute. Unpublished data.

193 Green, J. O.; Cowling, D. W. The nitrogen nutrition of grassland. *Proc. 8th int. Grassld Congr., Reading, 1960* 1961, pp. 126–9.

194 Greenhill, W. A. A study of the relative amounts of the protein and non-protein nitrogenous constituents occurring in pasture herbage, and their significance in the grazing of the herbage by stock. *Biochem. J.* 1936, **30,** 412–16.

195 Greenwood, D. J. Nitrogen transformations and the distribution of oxygen in soil. *Chemy Ind.* 1963, pp. 799–803.

196 Greenwood, E. A. N.; Goodall, D. W.; Titmanis, Z. V. Measurement of nitrogen deficiency in grass swards. *Pl. Soil* 1965, **23,** 97–116.

197 Greenwood, E. A. N.; Titmanis, Z. V. The effect of age on nitrogen stress and its relation to leaf nitrogen and leaf elongation in a grass. *Pl. Soil* 1966, **24,** 379–89.

198 Griffith, G. ap. Nitrogen and the nutritive value of grass. *Span* 1964, **7,** 18–20.

199 Griffith, G. ap; Jones, D. I. H.; Walters, R. J. K. Specific and varietal differences in sodium and potassium in grasses. *J. Sci. Fd Agric.* 1965, **16,** 94–8.

200 Griffith, W. K.; Teel, M. R. Effect of nitrogen and potassium fertilization, stubble height and clipping frequency on yield and persistence of orchardgrass. *Agron. J.* 1965, **57,** 147–9.

201 Griffith, W. K.; Teel, M .R.; Parker, H. E. Influence of nitrogen and potassium on the yield and chemical composition of orchardgrass. *Agron. J.* 1964, **56,** 473–5.

202 GRIMES, R. C.; WATKIN, B. R.; GALLAGHER, J. R. The growth of lambs grazing on perennial ryegrass, tall fescue and cocksfoot, with and without white clover, as related to the botanical and chemical composition of the pasture and pattern of fermentation in the rumen. *J. agric. Sci., Camb.* 1967, **68,** 11–21.

203 GROOT, T. DE. The influence of heavy nitrogen fertilization on the health of livestock. *J. Br. Grassld Soc.* 1963, **18,** 112–18.

204 GROOT, T. DE; KEUNING, J. A. Investigation into the effect of very high rates of nitrogen on the health and productivity of dairy cows at 'De Olde Weije', the Dutch nitrogenous fertilizer industry experimental farm at Vaassen. *Stikstof* (English edn) 1965, No. 9, 20–7.

205 HANAWALT, R. B. Environmental factors influencing the sorption of atmospheric ammonia by soils. *Proc. Soil Sci. Soc. Am.* 1969, **33**, 231–4.

206 HANCOCK, J. Grazing behaviour of cattle. *Anim. Breed. Abstr.* 1953, **21,** 1–13.

207 HARKESS, R. D. Studies in herbage digestibility. *J. Br. Grassld Soc.* 1963, **18,** 62–8.

208 HARMSEN, G. W.; KOLENBRANDER, G. J. Soil inorganic nitrogen. In: Soil Nitrogen. Agronomy, No. 10. Bartholomew, W. V.; Clark, F. E. [Eds] Madison, Wisconsin: Amer. Soc. Agron., 1965, pp. 43–92.

209 HART, M. L. T'. The influence of nitrogen fertilising on the botanical composition of grassland. *Stikstof* (English edn) 1957, No. 1, 33–7.

210 HARTMANS, J.; GRIFT, J. VAN DER. The effect of sulphur content in the feed on the copper status of cattle. *Jaarb. Inst. biol. scheik. Onderz. LandbGewass. 1964* 1964, pp. 145–55.

211 HAVRE, G. N.; DISHINGTON, I. E. The mineral composition of pasture as influenced by various types of heavy nitrogen dressings. *Acta Agric. scand.* 1962, **12,** 298–308.

212 HEAD, M. J.; ROOK, J. A. F. Hypomagnesaemia in dairy cattle and its possible relationship to ruminal ammonia production. *Nature, Lond.* 1955, **176,** 262–3.

213 HEDDLE, R. G. Nitrogenous fertilization of Italian ryegrass in spring. *J. Br. Grassld Soc.* 1968, **23,** 69–74.

214 HEMINGWAY, R. G. Magnesium, potassium, sodium and calcium contents of herbage as influenced by fertilizer treatments over a three-year period. *J. Br. Grassld Soc.* 1961, **16,** 106–16.

215 HEMINGWAY, R. G. Copper, molybdenum, manganese and iron contents of herbage as influenced by fertilizer treatments over a three-year period. *J. Br. Grassld Soc.* 1962, **17,** 182–7.

216 HEMINGWAY, R. G. Soil and herbage potassium levels in relation to yield. *J. Sci. Fd Agric.* 1963, **14,** 188–95.

217 HEMINGWAY, R. G.; INGLIS, J. S. S.; RITCHIE, N. S. Factors involved in hypomagnesaemia in sheep. *Proc. Br. Vet. Ass. Conf. on Hypomagnesaemia* 1960.

218 HENDERSON, R. The application of potassic fertilizers to pasture and the incidence of hypomagnesaemia. Tech. Series No. 1, Potash Ltd, London, 1960, pp. 23.

219 HENZELL, E. F. Nitrogen fixation and transfer by some tropical and temperate pasture legumes in sand culture. *Aust. J. exp. Agric. Anim. Husb.* 1962, **2,** 132–40.

220 HENZELL, E. F.; FERGUS, I. F.; MARTIN, A. E. Accumulation of soil nitrogen and carbon under a *Desmodium uncinatum* pasture. *Aust. J. exp. Agric. Anim. Husb.* 1966, **6,** 157–60.

221 HERRIOTT, J. B. D.; WELLS, D. A. Clover nitrogen and sward productivity. *J. Br. Grassld Soc.* 1960, **15,** 63–9.

222 HERRIOTT, J. B. D.; WELLS, D. A. Gülle as a grassland fertilizer. *J. Br. Grassld Soc.* 1962, **17,** 167–70.

223 HERRIOTT, J. B. D.; WELLS, D. A. The grazing animal and sward productivity. *J. agric. Sci., Camb.* 1963, **61,** 89–99.

224 HERRIOTT, J. B. D.; WELLS, D. A.; CROOKS, P. Gülle as a grassland fertilizer. Part 2. *J. Br. Grassld Soc.* 1963, **18,** 339–44.

225 HERRIOTT, J. B. D.; WELLS, D. A.; CROOKS, P. Gülle as a grassland fertilizer. Part 3. *J. Br. Grassld Soc.* 1965, **20,** 129–38.

226 HICKEY, F. Some metabolic aspects of the pasture/animal association. *N.Z. Jl agric. Res.* 1960, **3,** 468–84.

227 HODGSON, D. R.; DRAYCOTT, A. P. Aqueous ammonia compared with other nitrogenous fertilizers as solids and solutions on grass. *J. agric. Sci., Camb.* 1968, **71,** 195–203.

228 HODGSON, J.; SPEDDING, C. R. W. The health and performance of the grazing animal in relation to fertilizer nitrogen usage. 1. Calves. *J. agric. Sci., Camb.* 1966, **67,** 155–67.

229 HOLLIDAY, R.; WILMAN, D. The effect of white clover, fertilizer nitrogen and simulated animal residues on yield of grassland herbage. *J. Br. Grassld Soc.* 1962, **17,** 206–13.

230 HOLLIDAY, R.; WILMAN, D. The effect of fertilizer nitrogen and frequency of defoliation on yield of grassland herbage. *J. Br. Grassld Soc.* 1965, **20,** 32–40.

231 HOLMES, J. C.; LANG, R. W. Effects of fertilizer nitrogen and herbage dry-matter content on herbage intake and digestibility in bullocks. *Anim. Prod.* 1963, **5,** 17–26.

232 HOLMES, M. R. J. Evaluation of nitrogen fertilizers. In : Nitrogen and soil organic matter. *Tech. Bull. 15 Min. Agric., Fish., Fd* London: H.M.S.O., 1969, pp. 129–48.

233 Holmes, W. The intensive production of herbage for crop-drying. Part 2. A study of the effect of massive dressings of nitrogenous fertilizer and of the time of their application on the yield, chemical and botanical composition of two grass leys. *J. agric. Sci., Camb.* 1949, **39,** 128–41.

234 Holmes, W. The use of nitrogen in the management of pasture for cattle. *Herb. Abstr.* 1968, **38,** 265–77.

235 Holmes, W.; MacLusky, D. S. The intensive production of herbage for crop-drying. Part 5. The effect of continued massive applications of nitrogen with and without phosphate and potash on the yield of grassland herbage. *J. agric. Sci., Camb.* 1954, **45,** 129–39.

236 Holmes, W.; MacLusky, D. S. The intensive production of herbage for crop-drying. Part 6. A study of the effect of intensive nitrogen fertilizer treatment on species and strains of grass grown alone and with white clover. *J. agric. Sci., Camb.* 1955, **46**, 267–86.

237 Holt, E. C.; Fisher, F. L. Root development of Coastal Bermudagrass with high nitrogen fertilization. *Agron. J.* 1960, **52,** 593–6.

238 Hoogerkamp, M. [The effect of soil organic matter content on grassland productivity.] *Landbouwvoorlichting* 1965, **22,** 201–6. (Seen in *Herb. Abstr.* **36** : 23)

239 Hoogerkamp, M. The influence of organic matter in soil on grassland productivity. *Tech. Rep. 1 Grassld Res. Inst., Hurley* 1965, pp. 32–4.

240 Hoogerkamp, M.; Minderhoud, J. W. Herbage production of permanent grassland, resown grassland and leys, with special reference to the " years of depression ". *Proc. 10th int. Grassld Congr., Helsinki, 1966* 1966, pp. 282–7.

241 Hunt, I. V. Nitrogenous manuring for early bite, 1957. *Exp. Rec. 6 W. Scot. Agric. Coll.* 1963, pp. 34.

242 Hunt, I. V. The effect of utilisation of herbage on the response to fertilizer nitrogen. *Proc. 9th int. Grassld Congr., São Paulo, 1965* 1966, pp. 1113–9.

243 Hunt, I. V. The effect of age of sward on the yield and response of grass species to fertilizer nitrogen. *Proc. 10th int. Grassld Congr., Helsinki, 1966* 1966, pp. 249–54.

244 Hunt, L. A. Some implications of death and decay in pasture production. *J. Br. Grassld Soc.* 1965, **20,** 27–31.

245 Hvidsten, H. Studies on hypomagnesaemia in sheep as influenced by fertilizer treatment of pasture. *Z. Tierphysiol. Tierernähr. Futtermittelk.* 1967, **22,** 210–9.

246 Hylton, L. O.; Williams, D. E.; Ulrich, A.; Cornelius, D. R. Critical nitrate levels for growth of Italian ryegrass. *Crop Sci.* 1964, **4**, 16–19.

247 IMMINK, H. J.; GEURINK, J. H.; DEIJS, W. B. The determination of the higher fatty acids in grass and cow-faeces. *Jaarb. Inst. biol. scheik. Onderz. LandbGewass. 1965* 1965, pp. 103–7.

248 IMPERIAL CHEMICAL INDUSTRIES LTD. Jealott's Hill Research Station Guide to Field Experiments. 1957, pp. 26–8.

249 IMPERIAL CHEMICAL INDUSTRIES LTD. Jealott's Hill Research Station Guide to Field Experiments. 1966, p. 90.

250 IVINS, J. D. The relative palatability of herbage plants. *J. Br. Grassld Soc.* 1952, **7,** 43–54.

251 JAGTENBERG, W. D. [Can the best date for applying the first nitrogen to grassland be forecast ?] *Stikstof* 1966, **5,** No. 52, 216–22. (Seen in *Herb. Abstr.* **37** : 1064).

252 JAMESON, H. R. Liquid nitrogenous fertilizers. *J. agric. Sci., Camb.* 1959, **53,** 333–8.

253 JANSSON, S. L. Balance sheet and residual effects of fertilizer nitrogen in a 6-year study with N^{15}. *Soil Sci.* 1963, **95,** 31–7.

254 JARVIS, R. H. Studies on lucerne and lucerne-grass leys. 5. Plant population studies with lucerne. *J. agric. Sci., Camb.* 1962, **59,** 281–6.

255 JEATER, R. S. L. Liquefied ammonia in agriculture. *Agriculture, Lond.* 1966, **73,** 542–7.

256 JEATER, R. S. L. Comparisons of liquefied (anhydrous) ammonia and ammonium nitrate as nitrogenous fertilizers for grassland. *J. Br. Grassld Soc.* 1967, **22,** 225–9.

257 JENKINSON, D. S. Chemical tests for potentially available nitrogen in soil. *J. Sci. Fd Agric.* 1968, **19,** 160–8.

258 JENSEN, H. L. Nonsymbiotic nitrogen fixation. In : Soil Nitrogen. Agronomy, No. 10. Bartholomew, W. V. ; Clark, F. E. [Eds] Madison, Wisconsin : Amer. Soc. Agron., 1965, pp. 436–80.

259 JONES, D. I. H.; AP GRIFFITH, G.; WALTERS, R. J. K. The effect of nitrogen fertilizer on the water-soluble carbohydrate content of perennial ryegrass and cocksfoot. *J. Br. Grassld Soc.* 1961, **16,** 272–5.

260 JONES, D. I. H.; AP GRIFFITH, G.; WALTERS, R. J. K. The effect of nitrogen fertilizers on the water-soluble carbohydrate content of grasses. *J. agric. Sci., Camb.* 1965, **64,** 323–8.

261 JONES, D. J. C. The effect of sulphate of ammonia applications on the sulphur content of various grass and clover mixtures. *J. agric. Sci., Camb.* 1960, **54,** 188–94.

262 JONES, L. H. P.; HANDRECK, K. A. Silica in soils, plants and animals. *Adv. Agron.* 1967, **19,** 107–49.

263 JUDD, P. The effect of soil moisture regime and nitrogen application on the production of a perennial pasture mixture. *Proc. 9th int. Grassld Congr., São Paulo, 1965* 1966, pp. 1191–7.

264 KARRAKER, P. E.; BORTNER, C. E.; FERGUS, E. N. Nitrogen balance in lysimeters as affected by growing Kentucky bluegrass and certain legumes separately and together. *Bull. 557 Ky agric. Exp. Stn* 1950, pp. 16.

265 KEENEY, D. R.; BREMNER, J. M. A chemical index of soil nitrogen availability. *Nature, Lond.* 1966, **211**, 892–3.

266 KEMP, A. Hypomagnesaemia in milking cows : the response of serum magnesium to alterations in herbage composition resulting from potash and nitrogen dressings on pasture. *Neth. J. agric. Sci.* 1960, **8**, 281–304.

267 KEMP, A.; DEIJS, W. B.; KLUVERS, E. Influence of higher fatty acids on the availability of magnesium in milking cows. *Neth. J. agric. Sci.* 1966, **14**, 290–5.

268 KERSHAW, E. S. The crude protein and nitrate nitrogen relationship in S22 in response to nitrogen and potash fertilizer treatments. *J. Br. Grassld Soc.* 1963, **18**, 323–7.

269 KERSHAW, E. S.; BANTON, C. L. The mineral content of S22 ryegrass on calcareous loam soil in response to fertilizer treatments. *J. Sci. Fd Agric.* 1965, **16**, 698–701.

270 KILIAN, K. C.; ATTOE, O. J.; ENGELBERT, L. E. Urea-formaldehyde as a slowly available form of nitrogen for Kentucky bluegrass. *Agron. J.* 1966, **58**, 204–6.

271 KLETER, H. J. Influence of weather and nitrogen fertilization on white clover percentage of permanent grassland. *Neth. J. agric. Sci.* 1968, **16**, 43–52.

272 KNABE, O.; KNABE, B.; KREIL, W. [Effect of N application on the Cu content of pasture grass.] *Z. LandKult.* 1964, **5**, 245–58. (Seen in *Herb. Abstr.* **35** : 175).

273 KREIL, W.; WACKER, G.; KALTOFEN, H.; HEY, E. Heavy nitrogen fertilizing to pasture. *Proc. 9th int. Grassld Congr., São Paulo, 1965* 1966, pp. 1093–8.

274 KRESGE, C. B.; DECKER, A. M. Nutrient balance in Midland Bermudagrass as affected by differential nitrogen and potassium fertilization. 1. Forage yields and persistence. *Proc. 9th int. Grassld Congr., São Paulo, 1965* 1966, pp. 671–4.

275 KUNTZE, H. [Decomposition of grass and clover roots in soil.] *Z. Acker- u. PflBau* 1964, **120**, 383–400.

276 LAMBERT, D. A. The effect of the rate and timing of nitrogen application on the seed yield and components of yield on S48 timothy (*Phleum pratense* L.). *J. Br. Grassld Soc.* 1963, **18**, 154–7.

277 LAMBERT, D. A. The effect of level of nitrogen and cutting treatment on leaf area in swards of S48 timothy (*Phleum pratense* L.) and S215 meadow fescue (*Festuca pratensis* L.). *J. Br. Grassld Soc.* 1964, **19**, 396–402.

278 LANGER, R. H. M. Growth and nutrition of timothy (*Phleum pratense* L.). 4. The effect of nitrogen, phosphorus and potassium supply on growth during the first year. *Ann. appl. Biol.* 1959, **47**, 211–21.

279 LANGER, R. H. M. Growth and nutrition of timothy (*Phleum pratense* L.). 5. Growth and flowering at different levels of nitrogen. *Ann. appl. Biol.* 1959, **47,** 740–51.

280 LANGER, R. H. M. Tillering in herbage grasses. *Herb. Abstr.* 1963, **33,** 141–8.

281 LARGE, R. V. The effect of nitrogenous fertilizers on animal health. *N.A.A.S. q. Rev.* 1968, No. 79, 110–17.

282 LARGE, R. V.; SPEDDING, C. R. W. The health and performance of the grazing animal in relation to fertilizer nitrogen usage. 2. Weaned lambs. *J. agric. Sci., Camb.* 1966, **67,** 41–52.

283 LARSEN, S.; GUNARY, D. Ammonia loss from ammonical fertilizers applied to calcareous soils. *J. Sci. Fd Agric.* 1962, **13,** 566–72.

284 LARVOR, P.; GUÉGUEN, L. [Chemical composition of grass and grass tetany.] *Annls Zootech.* 1963, **12,** 39–52.

285 LAZENBY, A.; ROGERS, H. H. Selection criteria in grass breeding. 4. Effect of nitrogen and spacing on yield and its components. *J. agric. Sci., Camb.* 1965, **65,** 65–78.

286 LAZENBY, A.; ROGERS, H. H. Selection criteria in grass breeding. 5. Performance of *Lolium perenne* genotypes grown at different nitrogen levels and spacings. *J. agric. Sci., Camb.* 1965, **65,** 79–89.

287 LEGG, J. O.; ALLISON, F. E. Recovery of ^{15}N-tagged nitrogen from ammonium fixing soils. *Proc. Soil Sci. Soc. Am.* 1959, **23,** 131–4.

288 L'ESTRANGE, J. L.; OWEN, J. B.; WILMAN, D. The relationship between the serum magnesium concentration of grazing ewes and their dry matter intake and milk yield. *J. agric. Sci., Camb.* 1967, **68,** 165–71.

289 L'ESTRANGE, J. L.; OWEN, J. B.; WILMAN, D. Effects of a high level of nitrogenous fertilizer and date of cutting on the availability of the magnesium and calcium of herbage to sheep. *J. agric. Sci., Camb.* 1967, **68,** 173–8.

290 LEWIS, D. G. Partial inhibition of nitrate production by products of the mineralization of soil organic matter. *J. agric. Sci., Camb.* 1963, **61,** 349–52.

291 LINEHAN, P. A.; LOWE, J. Yielding capacity and grass/clover ratio of herbage swards as influenced by fertilizer treatments. *Proc. 8th int. Grassld Congr., Reading, 1960* 1961, pp. 133–7.

292 LONERAGAN, J. F. The nutrition of grasslands. In : Grasses and grasslands. C. Barnard [Ed.] London and Melbourne: Macmillan and Co. Ltd, 1964, pp. 206–20.

293 LOTERO, J.; WOODHOUSE, W. W.; PETERSEN, R. G. Local effect on fertility of urine voided by grazing cattle. *Agron. J.* 1966, **58,** 262–5.

294 LOW, A. J.; ARMITAGE, E. R. Irrigation of grassland. *J. agric. Sci., Camb.* 1959, **52,** 256–62.

295 Low, A. J.; Armitage, E. R. The composition of the leachate through cropped and uncropped soils in lysimeters compared with that of the rain. *Pl. Soil* (in press)

296 Low, A. J.; Piper, F. J. Urea as a fertilizer. Laboratory and pot-culture studies. *J. agric. Sci., Camb.* 1961, **57**, 249–55.

297 Lowe, J. Output of pastures under a clover nitrogen regime in Northern Ireland. *Proc. 10th int. Grassld Congr., Helsinki, 1966* 1966, pp. 187–91.

298 Lowe, J. Botanical development and output of a sward seeded with perennial ryegrass and white clover under stated fertilizer treatments. *Rec. agric. Res.* (*Nth. Ireld*) 1967, **16**(1), 75–91.

299 Lycklama, J. C. The absorption of ammonium and nitrate by perennial ryegrass. *Acta Bot. neerl.* 1963, **12**, 361–423.

300 Maass, G. [Experiments on the effect of ammonium and nitrate fertilizers on the mineral-nitrogen content of the soil.] *Z. PflErnähr.* [*Düng.*] *Bodenk.* 1961, **93,** 26–38.

301 Maass, G. [Experiments on the effect of ammonium and nitrate fertilizers on the mineral-nitrogen content of soil. 2. Pot experiments.] *Z. PflErnähr.* [*Düng.*] *Bodenk.* 1962, **98,** 146–54.

302 McAllister, J. S. V. A study of the responses to alternative sources of nitrogen in Northern Ireland. 1. Effects of time of application on yield and mineral composition of Italian ryegrass. *Rec. agric. Res.* (*Nth. Ireld*) 1966, **15**(2), 67–87.

303 McAllister, J. S. V. A study of the responses to alternative sources of nitrogen in Northern Ireland. 2. Effects of moderate and of heavy dressings on the yield and mineral composition of Italian ryegrass. *Rec. agric. Res.* (*Nth. Ireld*) 1966, **15**(2), 89–110.

304 McAllister, J. S. V. A study of the responses to alternative sources of nitrogen in Northern Ireland. 4. Effects of moderate and heavy dressings on the nitrate content of the surface soils. *Rec. agric. Res.* (*Nth. Ireld*) 1967, **16**(1), 41–55.

305 McAllister, J. S. V.; McConaghy, S. The application of heavy dressings of nitrogen to pasture. *Res. exp. Rec. Minist. Agric. Nth. Ire.* 1960, **10,** 87–104.

306 McAllister, J. S. V.; McConaghy, S.; Coey, W. E.; Kerr, J. A. M. The effects of different nitrogen treatments on the spring growth of ryegrass. 1. Effects on yield and nitrogen content of the herbage. *Rec. agric. Res.* (*Nth. Ireld*) 1965, **14**(2), 15–29.

307 McAuliffe, C.; Chamblee, D. S.; Uribe-Arango, H.; Woodhouse, W. W. Influence of inorganic nitrogen on nitrogen fixation by legumes as revealed by N^{15}. *Agron. J.* 1958, **50,** 334–7.

308 McCarrick, R. B.; Wilson, R. K. Effects of nitrogen fertilization of mixed swards on herbage yield, dry matter digestibility and voluntary food intake of the conserved herbages. *J. Br. Grassld Soc.* 1966, **21,** 195–9.

309 McConaghy, S.; Stewart, J. W. B.; Lowe, J. The effect on soils and herbage of a nitrogenous fertilizer containing ammonium nitrate, applied regularly at varying levels. *Res. exp. Rec. Minist. Agric. Nth. Ire.* 1962, **12,** 71–92.

310 MacFarlan, J. Time of application of nitrogen as a factor influencing the yield of herbage on permanent pasture. *Exp. Agric.* 1939, **7,** 155–61.

311 McKee, H. S. Nitrogen metabolism in plants. Oxford : Clarendon Press, 1962, pp. 728.

312 McLaren, G. A. Symposium on microbial digestion in ruminants: nitrogen metabolism in the rumen. *J. Anim. Sci.* 1964, **23,** 577–90.

313 McLean, E. O.; Adams, D.; Franklin, R. E. Cation exchange capacities of plant roots as related to their nitrogen contents. *Proc. Soil Sci. Soc. Am.* 1956, **20,** 345–7.

314 McLean, J. W.; Thomson, G. G.; Iversen, C. E.; Jagusch, K. T.; Lawson, B. M. Sheep production and health on pure species pastures. *Proc. 24th Conf. N.Z. Grassld Ass.* 1962, pp. 57–70.

315 MacLeod, L. B. Effect of nitrogen and potassium on the yield, botanical composition and competition for nutrients in three alfalfa-grass associations. *Agron. J.* 1965, **57,** 129–34.

316 MacLusky, D. S. Some estimates of the areas of pasture fouled by the excreta of dairy cows. *J. Br. Grassld Soc.* 1960, **15,** 181–8.

317 MacLusky, D. S.; Morris, D. W. Grazing methods, stocking rate and grassland production. *Agric. Prog.* 1964, **39,** 97–108.

318 McNaught, K. J. Potassium deficiency in pastures. 1. Potassium content of legumes and grasses. *N.Z. Jl agric. Res.* 1958, **1,** 148–81.

319 Mahoney, A. W.; Poulton, B. R. Effects of nitrogen fertilization and date of harvest on the acceptibility of timothy forage. *J. Dairy Sci.* 1962, **45,** 1575.

320 Malo, B. A.; Purvis, E. R. Soil absorption of atmospheric ammonia. *Soil Sci.* 1964, **97,** 242–7.

321 Marten, G. C.; Donker, J. D. Selective grazing induced by animal excreta. 2. Investigation of a causal theory. *J. Dairy Sci.* 1964, **47,** 871–4.

322 Martin, A. E. Soil chemistry and the ‘ nitrogen cycle ’. *J. Aust. Inst. agric. Sci.* 1966, **32,** 145–6.

323 Martin, A. E.; Henzell, E. F.; Ross, P. J.; Haydock, K. P. Isotopic studies on the uptake of nitrogen by pasture grasses. 1. Recovery of fertilizer nitrogen from the soil : plant system using Rhodes grass in pots. *Aust. J. Soil Res.* 1963, **1**, 169–84.

324 Martin, A. E.; Skyring, G. W. Losses of nitrogen from soil/plant system. In : A review of nitrogen in the tropics. with particular reference to pastures. A symposium. *Bull. 46 Commonw. Bur. Past. Fld Crops* Farnham Royal : Commonw. Agric. Bur., 1962, pp. 19–34.

325 Martin, T. W. The role of white clover in grassland. *Herb. Abstr.* 1960, **30**, 159–64.

326 Mazurak, A. P.; Conard, E. C. Changes in content of total nitrogen and organic matter in three Nebraska soils after seven years of cropping treatments. *Agron. J.* 1966, **58**, 85–8.

327 Metson, A. J.; Hurst, F. B. Effects of sheep dung and urine on a soil under pasture at Lincoln, Canterbury, with particular reference to potassium and nitrogen equilibria. *N.Z. Jl Sci. Technol. Sect. A* 1953, **35**, 327–59.

328 Metson, A. J.; Saunders, W. M. H.; Collie, T. W. ; Graham, V. W. Chemical composition of pastures in relation to grass tetany in beef breeding cows. *N.Z. Jl agric. Res.* 1966, **9**, 410–36.

329 Miles, D. G.; Williams, I. G. Winter hardiness in pasture varieties. *Rep. Welsh Pl. Breed. Stn 1963* 1964, pp. 70–1.

330 Miller, R. H.; Schmidt, E. L. Uptake and assimilation of amino acids supplied to the sterile soil : root environment of the bean plant (*Phaseolus vulgaris*). *Soil Sci.* 1965, **100,** 323–30.

331 Miller, W. J.; Adams, W. E.; Nussbaumer, R.; McCreery, R. A.; Perkins, H. F. Zinc content of Coastal Bermudagrass as influenced by frequency and season of harvest, location, and level of N and lime. *Agron. J.* 1964, **56,** 198–201.

332 Ministry of Agriculture, Fisheries and Food, England and Wales. Grass and grassland. *Bull. 154 Min. Agric., Fish., Fd* 1966, 4th Edn, pp. 114.

333 Minson, D. J.; Raymond, W. F.; Harris, C. E. Studies in the digestibility of herbage. 8. The digestibility of S37 cocksfoot, S23 ryegrass and S24 ryegrass. *J. Br. Grassld Soc.* 1960, **15**, 174–80.

334 Mitchell, W. H. Influence of nitrogen and irrigation on the root and top growth of forage crops. *Bull. 341 Delaware agric. Exp. Stn* 1962, pp. 32.

335 Moir, R. J. Personal communication, 1966. [Cited by Allaway, W. H.; Thompson, J. F. Sulfur in the nutrition of plants and animals. *Soil Sci.* 1966, **101**, 240–7]

336 MOLEN, H. VAN DER. Experience with high amounts of nitrogen on grassland at nitrogen experimental farms in the Netherlands. *J. Br. Grassld Soc.* 1963, **18,** 235–42.

337 MOLEN, H. VAN DER. Hypomagnesaemia and grass fertilization in the Netherlands. *Outl. Agric.* 1964, **4,** 55–63.

338 MOLEN, H. VAN DER. Nitrogen usage in Western Europe. *Stikstof* (English edn) 1966, No. 10, 3–7.

339 MOLONEY, D.; MURPHY, W. E. The effect of different levels of N on a grass clover sward under grazing conditions. *Ir. J. agric. Res.* 1963, **2,** 1–12.

340 MOON, F. E. The influence of manurial treatment on the carotene content of poor pasture grass and on the relationship of this constituent to the ash and organic fractions. *J. agric. Sci., Camb.* 1939, **29,** 524–43.

341 MOORE, A. W. Non-symbiotic nitrogen fixation in soil and soil-plant systems. *Soils Fertil., Harpenden* 1966, **29,** 113–28.

342 MORTENSEN, W. P.; BAKER, A. S.; DERMANIS, P. Effects of cutting frequency of orchardgrass and nitrogen rate on yield, plant nutrient composition and removal. *Agron. J.* 1964, **56,** 316–20.

343 MOUAT, M. C. H.; WALKER, T. W. Competition for nutrients between grasses and white clover. 1. Effect of grass species and nitrogen supply. *Pl. Soil* 1959, **11,** 30–40.

344 MUDD, C. H.; MEADOWCROFT, S. C. Comparison between the improvement of pastures by the use of fertilizers and by reseeding. *Expl Husb.* 1964, **10,** 66–84.

345 MULDER, E. G. Fertilizer *vs.* legume nitrogen for grasslands. *Proc. 6th int. Grassld Congr., Pasadena, 1952* 1952, pp. 740–8.

346 MULLALY, J. V.; MCPHERSON, J. B.; MANN, A. P.; ROONEY, D. R. The effect of length of legume and non-legume leys on gravimetric soil nitrogen at some locations in the Victorian wheat areas. *Aust. J. exp. Agric. Anim. Husb.* 1967, **7,** 568–71.

347 MUNRO, I. A. Irrigation of grassland. The influence of irrigation and nitrogen treatments on the yield and utilization of a riverside meadow. *J. Br. Grassld Soc.* 1958, **13,** 213–21.

348 MUNRO, P. E. Inhibition of nitrifiers by grass root extracts. *J. appl. Ecol.* 1966, **3,** 231–8.

349 NATIONAL AGRICULTURAL ADVISORY SERVICE. Fertilizer recommendations for agricultural and horticultural crops. *Advis. Pap. 4 N.A.A.S., Minist. Agric., Fish., Fd* 1967.

350 NATIONAL INSTITUTE FOR RESEARCH IN DAIRYING. *Annual Report, 1954* 1954, pp. 36–7.

351 NEENAN, M. *Proc. 8th int. Grassld Congr., Reading, 1960* 1961, p. 153.

352 NEWMAN, R. J.; ALLEN, B. F.; COOK, M. G. The effect of nitrogen on winter pasture production in southern Victoria. *Aust. J. exp. Agric. Anim. Husb.* 1962, **2,** 20–4.

353 New Zealand Department For Scientific and Industrial Research. Plant Chemistry Division. *Quadrennial Rep. 1962–65* 1966, pp. 32–3.

354 Nielsen, B. F. Plant production, transpiration ratio and nutrient ratios as influenced by interactions between water and nitrogen. Thesis, R. Vet. Agric. Coll., Copenhagen, 1963.

355 Nielsen, K. F.; Cunningham, R. K. The effects of soil temperature and form and level of nitrogen on growth and chemical composition of Italian ryegrass. *Proc. Soil Sci. Soc. Am.* 1964, **28**, 213–18.

356 Nommik, H. Investigations on denitrification in soil. *Acta Agric. scand.* 1956, **6**, 195–228.

357 Nommik, H. Ammonium fixation and other reactions involving a non-enzymatic immobilization of mineral nitrogen in soils. In : Soil Nitrogen. Agronomy, No. 10. Bartholomew, W. V.; Clark, F. E. [Eds] Madison, Wisconsin : Amer. Soc. Agron., 1965, pp. 193–258.

358 Nommik, H.; Nilsson, K. O. Nitrification and movement of anhydrous ammonia in soil. *Acta Agric. scand.* 1963, **13**, 205–19.

359 Norman, M. J. T. Intervals of superphosphate application to downland permanent pasture. *J. agric. Sci., Camb.* 1956, **47**, 157–71.

360 North of Scotland College of Agriculture. Grassland Experimental Centre, Muchalls, Kincardine. Guide to Experiments. 1965, pp. 6–7.

361 Nowakowski, T. Z. The effect of different nitrogenous fertilizers, applied as solids or solutions, on the yield and nitrate-N content of established grass and newly-sown ryegrass. *J. agric. Sci., Camb.* 1961, **56**, 287–92.

362 Nowakowski, T. Z. Effects of nitrogen fertilizers on total nitrogen, soluble nitrogen and soluble carbohydrates contents of grass. *J. agric. Sci., Camb.* 1962, **59**, 387–92.

363 Nowakowski, T. Z. Mineral fertilisation and organic composition of herbage. *Proc. 2nd Reg. Conf. int. Potash Inst., Morat, Switzerland, 1964* 1964, pp. 63–73.

364 Nowakowski, T. Z.; Cunningham, R. K. Nitrogen fractions and soluble carbohydrates in Italian ryegrass. 2. Effects of light intensity, form and level of nitrogen. *J. Sci. Fd Agric.* 1966, **17**, 145–50.

365 Nowakowski, T. Z.; Cunningham, R. K.; Nielsen, K. F. Nitrogen fractions and soluble carbohydrates in Italian ryegrass. 1. Effects of soil temperature, form and level of nitrogen. *J. Sci. Fd Agric.* 1965, **16**, 124–34.

366 Nowakowski, T. Z.; Gasser, J. K. R. The effect of a nitrification inhibitor on the concentration of nitrate in plants. *J. agric. Sci., Camb.* 1967, **68**, 131–3.

367 O'Brien, T. A. The influence of nitrogen on seedling and early growth of perennial ryegrass and cocksfoot. *N.Z. Jl agric. Res.* 1960, **3,** 399–411.

368 Odhuba, E. K.; Reid, R. L.; Jung, G. A. Nutritive evaluation of tall fescue pasture. *J. Anim. Sci.* 1965, **24,** 1216.

369 Oostendorp, D. [Nitrogen manuring and grass growth in spring.] *Landbouwvoorlichting* 1965, **22,** 29–35. (Seen in *Herb. Abstr.* **36** : 16)

370 Oswalt, D. L.; Bertrand, A. R.; Teel, M. R. Influence of nitrogen fertilization and clipping on grass roots. *Proc. Soil Sci. Soc. Am.* 1959, **23,** 228–30.

371 Oxford, A. E. A guide to rumen microbiology. *Bull. 160 Dep. scient. ind. Res. N.Z.* 1964, pp. 103.

372 Padmos, L. Nitrogen fertilization in the Netherlands. *Stikstof* (English edn) 1966, No. 10, 56–65.

373 Parker, C. A. Non-symbiotic nitrogen-fixing bacteria in soil. 3. Total nitrogen changes in a field soil. *J. Soil Sci.* 1957, **8,** 48–59.

374 Parks, W. L.; Fisher, W. B. Influence of soil temperature and nitrogen on ryegrass growth and chemical composition. *Proc. Soil Sci. Soc. Am.* 1958, **22,** 257–9.

375 Parr, J. F. Biochemical considerations for increasing the efficiency of nitrogen fertilizers. *Soils Fertil., Harpenden* 1967, **30,** 207–13.

376 Patrick, W. H.; Wyatt, R. Soil nitrogen loss as a result of alternate submergence and drying. *Proc. Soil Sci. Soc. Am.* 1964, **28,** 647–53.

377 Paul, E. A.; Schmidt, E. L. Formation of free amino acids in rhizosphere and non-rhizosphere soil. *Proc. Soil Sci. Soc. Am.* 1961, **25,** 359–62.

378 Penman, H. L. Woburn irrigation, 1951–59. 2. Results for grass. *J. agric. Sci., Camb.* 1962, **58,** 349–64.

379 Perez, C. B.; Story, C. D. The effect of nitrate in nitrogen-fertilized hays on fermentation *in vitro*. *J. Anim. Sci.* 1960, **19,** 1311.

380 Petersen, R. G.; Woodhouse, W. W.; Lucas, H. L. The distribution of excreta by freely grazing cattle and its effect on pasture fertility. 2. Effect of returned excreta on the residual concentration of some fertilizer elements. *Agron. J.* 1956, **48,** 444–9.

381 Power, J. F. Mineralization of nitrogen in grass roots. *Proc. Soil Sci. Soc. Am.* 1968, **32,** 673–4.

382 Prine, G. M.; Gardner, F. P.; Willard, C. J. Irrigation and nitrogen treatment of forage crops. *Res. Circ. 119 Ohio agric. Exp. Stn* 1963, pp. 35.

383 RAE, A. L.; BROUGHAM, R. W.; GLENDAY, A. C.; BUTLER, G. W. Pasture type in relation to live-weight gain, carcass composition, iodine nutrition and some rumen characteristics of sheep. 1. Live-weight growth of the sheep. *J. agric. Sci., Camb.* 1963, **61,** 187–90.

384 RAHMAN, H.; McDONALD, P.; SIMPSON, K. Effect of nitrogen and potassium fertilizers on the mineral status of perennial ryegrass. 1. Mineral content. *J. Sci. Fd Agric.* 1960, **11,** 422–8.

385 RAININKO, K. The effects of nitrogen fertilization, irrigation and number of harvestings upon leys established with various seed mixtures. *Suom. maatal. Seur. Julk.* No. 112, pp 137.

386 RAMAGE, C. H.; EBY, C.; MATHER, R. E.; PURVIS, E. R. Yield and chemical compositions of grasses fertilized heavily with nitrogen. *Agron. J.* 1958, **50,** 59–62.

387 RAYMOND, W. F.; SPEDDING, C. R. W. The effect of fertilizers on the nutritive value and production potential of forages. *Proc. Fertil. Soc.* 1965, **88,** pp. 34.

388 RAYMOND, W. F.; SPEDDING, C. R. W. Nitrogenous fertilizers and the feed value of grass. *Proc. 1st Gen. Meet. Eur. Grassld Fed.* 1965, pp. 151–60.

389 REID, D. The response of herbage yields and quality to a wide range of nitrogen application rates. *Proc. 10th int. Grassld Congr., Helsinki, 1966* 1966, pp. 209–13.

390 REID, D.; CASTLE, M. E. The response of grass-clover and pure grass leys to irrigation and fertilizer nitrogen treatment. 1. Irrigation effects. *J. agric. Sci., Camb.* 1965, **64,** 185–94.

391 REID, D.; CASTLE, M. E. The response of grass-clover and pure grass leys to irrigation and fertilizer nitrogen treatment. 2. Clover and fertilizer nitrogen effects. *J. agric. Sci. Camb.* 1965, **65,** 109–19.

392 REID, R. L.; JUNG, G. A. The influence of fertilizer treatment on the intake, digestibility and palatability of tall fescue hay. *J. Anim. Sci.* 1965, **24,** 615–25.

393 REID, R. L.; JUNG, G. A.; KINSEY, C. M. Nutritive value of nitrogen-fertilized orchardgrass pasture at different periods of the year. *Agron. J.* 1967, **59,** 519–25.

394 REID, R. L.; JUNG, G. A.; MURRAY, S. J. Nitrogen fertilization in relation to the palatability and nutritive value of orchardgrass. *J. Anim. Sci.* 1966, **25,** 636–45.

395 REITH, J. W. S.; INKSON, R. H. E. and collaborators. The effects of fertilizers on herbage production. 1. The effect of nitrogen, phosphate and potash on yield. *J. agric. Sci., Camb.* 1961, **56,** 17–29.

396 Reith, J. W. S.; Inkson, R. H. E. and collaborators. The effects of fertilizers on herbage production. 2. The effect of nitrogen, phosphorus and potassium on botanical and chemical composition. *J. agric. Sci., Camb.* 1964, **63,** 209–19.

397 Reith, J. W. S.; Mitchell, R. L. The effect of soil treatment on trace element uptake by plants. In : Plant analysis and fertilizer problems. Vol. 4. Bould, C.; Prevot, P.; Magness, J. R. [Eds] E. Lansing, Michigan : Amer. Soc. Hort. Sci., 1964, pp. 241–54.

398 Rhykerd, C. L.; Dillon, J. E.; Noller, C. H.; Burns, J. C. The influence of nitrogen fertilization and drying method on yield and chemical composition of *Dactylis glomerata*, *Bromus inermis* and *Phleum pratense*. *Proc. 10th int. Grassld Congr., Helsinki, 1966* 1966, pp. 214–18.

399 Richardson, H. L. The nitrogen cycle in grassland with special reference to the Rothamsted Park grass experiment. *J. agric. Sci., Camb.* 1938, **28,** 73–121.

400 Robinson, J. B. Nitrification in a New Zealand grassland soil. *Pl. Soil* 1963, **19,** 173–83.

401 Robinson, R. R.; Sprague, V. G. The clover populations and yields of a Kentucky bluegrass sod as affected by nitrogen fertilization, clipping treatments and irrigation. *Agron. J.* 1947, **39,** 107–16.

402 Russel, J. S. Soil fertility changes in the long-term experimental plots at Kybybolite, South Australia. 1. Changes in pH, total nitrogen, organic carbon and bulk density. *Aust. J. agric. Res.* 1960, **11,** 902–26.

403 Russell, E. W. Soil conditions and plant growth. London : Longmans, Green & Co., 9th edn, 1961, p. 350.

404 Ryle, G. J. A. A comparison of leaf and tiller growth in seven perennial grasses as influenced by nitrogen and temperature. *J. Br. Grassld Soc.* 1964, **19,** 281–90.

405 Ryle, G. J. A. Effects of two levels of applied nitrogen on the growth of S37 cocksfoot in small simulated swards in a controlled environment. *J. Br. Grassld Soc.* 1970, **25,** 20–9.

406 Schmidt, D. R.; Tenpas, G. H. Seasonal response of grasses fertilized with nitrogen compared to a legume-grass mixture. *Agron. J.* 1965, **57,** 428–31.

407 Schöllhorn, J. [The effect on grassland of gülle [liquid manure] which has been stored for varying periods and at different dilutions.] *Z. Acker-u. Pflbau* 1955, **100,** 211–38.

408 Schreven, D. A. van. Some factors affecting the uptake of nitrogen by legumes. In : Nutrition of the legumes. Hallsworth, E. G. [Ed.] London : Butterworth Sci. Publ., 1958, pp. 137–63.

409 Schreven, D. A. van. [Some microbiological aspects of nitrogen metabolism in soil.] *Annls Inst. Pasteur, Paris* 1965, Suppl. to No. 3, 19–49.

410 SEARS, P. D. Soil fertility and pasture growth. *J. Br. Grassld Soc.* 1950, **5**, 267–80.

411 SEARS, P. D. Pasture growth and soil fertility. 1. The influence of red and white clovers, superphosphate, lime and sheep grazing on pasture yields and botanical composition. *N.Z. Jl Sci. Technol. Sect. A* 1953, **35**, Suppl., 1–29.

412 SEARS, P. D. Pasture growth and soil fertility. 7. General discussion of the experimental results and of their application to farming practice in New Zealand. *N.Z. Jl Sci. Technol. Sect. A* 1953, **35**, 221–36.

413 SEARS, P. D. Grass/clover relationships in New Zealand. *Proc. 8th int Grassld Congr., Reading, 1960* 1961, pp. 130–3.

414 SEARS, P. D.; EVANS, L. T. Pasture growth and soil fertility. 3. The influence of red and white clovers, superphosphate, lime and dung and urine on soil composition and on earthworm and grass-grub populations. *N.Z. Jl Sci. Technol. Sect. A* 1953, **35**, Suppl., 42–52.

415 SEARS, P. D.; GOODALL, V. C.; JACKMAN, R. H.; ROBINSON, G. S. Pasture growth and soil fertility. 8. The influence of grasses, white clover, fertilizers and the return of herbage clippings on pasture production of an impoverished soil. *N.Z. Jl agric. Res.* 1965, **8**, 270–83.

416 SEARS, P. D.; GOODALL, V. C.; NEWBOLD, R. P. The effect of sheep droppings on yield, botanical composition and chemical composition of pasture. 2. Results for the years 1942–44 and final summary of the trial. *N.Z. Jl Sci. Technol. Sect. A* 1948, **30**, 231–50.

417 SEARS, P. D.; NEWBOLD, R. P. The effect of sheep droppings on yield, botanical composition and chemical composition of pasture. 1. *N.Z. Jl Sci. Technol. Sect. A* 1942, **24**, 36–61.

418 SEARS, P. D.; THURSTON, W. G. Effect of sheep droppings on yield, botanical composition and chemical composition of pasture. 3. Results of field trial at Lincoln, Canterbury, for the years 1944–1947. *N.Z. Jl Sci. Technol. Sect. A* 1953, **34**, 445–59.

419 SHAW, P. G.; BROCKMAN, J. S. North Wyke Experiment Station, unpublished data.

420 SHAW, P. G.; BROCKMAN, J. S.; WOLTON, K. M. The effect of cutting and grazing on the response of grass/white-clover swards to fertilizer nitrogen. *Proc. 10th int. Grassld Congr., Helsinki, 1966* 1966, pp. 240–4.

421 SIBMA, L. Regrowth of grass. *Jaarb. Inst. biol. scheik. Onderz. LandbGewass.* 1966, pp. 67–72.

422 SIMON, W.; EICH, D.; ZAJONZ, A. [Preliminary report on the relationship between quantity of roots and the value as a preceding crop of several leguminous and grass species when grown as a main crop on light soils.] *Z. Acker-u. PflBau* 1957, **104**, 71–88.

423 SIMPSON, J. R. Mineral nitrogen fluctuations in soils under improved pasture in southern New South Wales. *Aust. J. agric. Res.* 1962, **13,** 1059–72.

424 SIMPSON, J. R. The transference of nitrogen from pasture legumes to an associated grass under several systems of management in pot culture. *Aust. J. agric. Res.* 1965, **16,** 915–26.

425 SIMPSON, J. R.; FRENEY, J. R. The fate of labelled mineral nitrogen after addition to three pasture soils of different organic matter contents. *Aust. J. agric. Res.* 1967, **18,** 613–23.

426 SINCLAIR, K. B.; JONES, D. I. H. Nitrate poisoning in ruminants. *Rep. Welsh Pl. Breed. Stn 1962* 1963, pp. 97–101.

427 SMITH, A. M.; WANG, T. The carotene content of certain species of grassland herbage. *J. agric. Sci., Camb.* 1941, **31,** 370–8.

428 SMITH, D.; JEWISS, O. R. Effects of temperature and nitrogen supply on the growth of timothy (*Phleum pratense* L.). *Ann. appl. Biol.* 1966, **58,** 145–57.

429 SMITH, D.; SUND, J. M. Influence of stage of growth and soil nitrogen on nitrate content of herbage of alfalfa, red clover, ladino clover, trefoil and bromegrass. *J. agric. Fd Chem.* 1965, **13,** 81–4.

430 SONNEVELD, A. Beef production on intensively managed grassland. *Jaarb. Inst. biol. scheik. Onderz. LandbGewass.* 1959, p. 53.

431 SOULIDES, D. A.; CLARK, F. E. Nitrification in grassland soils. *Proc. Soil Sci. Soc. Am.* 1958, **22,** 308–11.

432 SPRAGUE, V. G.; GARBER, R. J. Effect of time and height of cutting and nitrogen fertilization on the persistence of the legume and production of orchard grass-Ladino and brome grass-Ladino associations. *Agron. J.* 1950, **42,** 586–93.

433 STERN, W. R.; DONALD, C. M. Light relationships on grass-clover swards. *Aust. J. agric. Res.* 1962, **13,** 599–614.

434 STERN, W. R.; DONALD, C. M. The influence of leaf area and radiation on the growth of clover in swards. *Aust. J. agric. Res.* 1962, **13,** 615–23.

435 STEVENSON, C. M. An analysis of the chemical composition of rain-water and air over the British Isles and Eire for the years 1959–64. *Q. Jl R. met. Soc.* 1968, **94,** 56–70.

436 STEVENSON, F. J. Origin and distribution of nitrogen in soil. In : Soil nitrogen. Agronomy, No. 10. Bartholomew, W. V.; Clark, F. E. [Eds] Madison, Wisconsin : Amer. Soc. Agron., 1965, pp. 1–42.

437 STEWART, A. B.; HOLMES, W. Nitrogenous manuring of grassland. 1. Some effects of heavy dressings of nitrogen on the mineral composition. *J. Sci. Fd Agric.* 1953, **4,** 401–8.

438 STILES, W. Ten years of irrigation experiments. *A. Rep. Grassld Res. Inst. Hurley 1965* 1966, pp. 57–66.

439 STILES, W.; WILLIAMS, T. E. The response of a ryegrass/white clover sward to various irrigation regimes. *J. agric. Sci., Camb.* 1965, **65,** 351–64.

440 STILLINGS, B. R.; BRATZLER, J. W.; MARRIOTT, L. F.; MILLER, R. C. Utilization of magnesium and other minerals by ruminants consuming low and high nitrogen-containing forages and vitamin D. *J. Anim. Sci.* 1964, **23,** 1148–54.

441 STRONG, T. H.; TRUMBLE, H. C. Excretion of nitrogen by leguminous plants. *Nature, Lond.* 1939, **143,** 286.

442 SULLIVAN, J. T.; SPRAGUE, V. G. Composition of the roots and stubble of perennial ryegrass following partial defoliation. *Pl. Physiol., Lancaster* 1943, **18,** 656–70.

443 TAYLER, J. C.; RUDMAN, J. E. The production of fattening cattle and extension of autumn grazing following three rates of application of nitrogenous fertilizer to a ryegrass/white clover sward. *J. agric. Sci., Camb.* 1960, **55,** 75–90.

444 TAYLER, R. S. The irrigation of grassland. *Outl. Agric.* 1965, **4,** 234–42.

445 TEMPLEMAN, W. G. Urea as a fertilizer. *J. agric. Sci., Camb.* 1961, **57,** 237–9.

446 TEMPLETON, W. C.; TAYLOR, T. H. Some effects of nitrogen, phosphorus and potassium fertilization on botanical composition of a tall fescue—white clover sward. *Agron. J.* 1966, **58,** 569–72.

447 THERON, J. J. Influence of plants on the mineralization of nitrogen and the maintenance of organic matter in the soil. *J. agric. Sci., Camb.* 1951, **41,** 289–96.

448 THERON, J. J. The mineralization of nitrogen in soils under grass. *S. Afr. J. agric. Sci.* 1963, **6,** 155–64.

449 THERON, J. J. The influence of fertilizers on the organic matter content of the soil under natural veld. *S. Afr. J. agric. Sci.* 1965, **8,** 525–34.

450 THERON, J. J.; HAYLETT, D. G. The regeneration of soil humus under a grass ley. *Exp. Agric.* 1953, **21,** 86–98.

451 THOMAS, P. T. Presidential address. *J. Br. Grassld Soc.* 1966, **21,** 1–6.

452 TOOMRE, R. I. Effect of high rates of nitrogenous fertilizers on the yield and chemical composition of grasses. *Proc. 10th int. Grassld Congr., Helsinki, 1966* 1966, pp. 227–30.

453 TROUGHTON, A. The underground organs of herbage grasses. *Bull. 44 Commonw. Bur. Past. Fld Crops* Farnham Royal: Commw. Agric. Bur., 1957, pp. 163.

454 TUIL, H. D. W. VAN. Organic salts in plants in relation to nutrition and growth. *Agric. Res. Reps* [*Versl. landbouwk. Onderz.*] 1965, No. 657, pp. 83.

455 TUKEY, H. B. J.; MORGAN, T. V. The occurrence of leaching from above ground plant parts and the nature of the material leached. *Proc. 16th int. Hort. Congr., 1962* 1964, **4,** 153–60.

456 TURNER, R. *et al.* The use of liquid manure on farms. [*Publ.*] *Dep. Agric. Scotland* Edinburgh: H.M.S.O., 1958, pp. 32.

457 VERCOE, J. M. Some observations on the nitrogen and energy losses in the faeces and urine of grazing sheep. *Proc. Aust. Soc. Anim. Prod.* 1962, **4,** 160–2.

458 VEZ, A. [Study of the development of white clover in relation to fluctuations in certain organic substances.] *Ber. schweiz. bot. Ges.* 1961, **71,** 118–72.

459 VIETS, F. G. The plant's need for and use of nitrogen. In : Soil Nitrogen. Agronomy, No. 10. Bartholomew, W. V. ; Clark, F. E. [Eds] Madison, Wisconsin : Amer. Soc. Agron., 1965, pp. 503–49.

460 VINCENT, J. M. Environmental factors in the fixation of nitrogen by the legume. In : Soil Nitrogen. Agronomy, No. 10. Bartholomew, W. V.; Clark, F. E. [Eds] Madison, Wisconsin : Amer. Soc. Agron., 1965, pp. 384–435.

461 VIRTANEN, A. I.; MIETTINEN, J. K. Biological nitrogen fixation. In : Plant physiology. Vol. 3. Steward, F. C. [Ed.] New York and London : Academic Press, 1963, pp. 539–668.

462 VOLK, G. M. Volatile loss of ammonia following surface application of urea to turf or bare soils. *Agron. J.* 1959, **51,** 746–9.

463 VOSE, P. B. The physiology of the vegetative grass plant. *Rep. Welsh Pl. Breed. Stn 1959* 1960, pp. 17–18.

464 WAGNER, R. E. Influence of legume and fertilizer nitrogen on forage production and botanical composition. *Agron. J.* 1954, **46,** 167–71.

465 WAGNER, R. E. Legume nitrogen versus fertilizer nitrogen in protein production of forage. *Agron. J.* 1954, **46,** 233–7.

466 WAITE, R. The chemical composition of grasses in relation to agronomical practice. *Proc. Nutr. Soc.* 1965, **24,** 38–46.

467 WALKER, T. W. The nitrogen cycle in grassland soils. *J. Sci. Fd Agric.* 1956, **7,** 66–72.

468 WALKER, T. W. Problems of soil fertility in a grass-animal regime. *Trans. int. Soil Sci. Conf., New Zealand* 1962, pp. 704–14.

469 WALKER, T. W. The significance of phosphorus in pedogenesis. In : Experimental pedology. Hallsworth, E. G.; Crawford, D. V. [Eds] London : Butterworths, 1965, pp. 295–316.

470 WALKER, T. W.; ADAMS, A. F. R. Competition for sulphur in a grass-clover association. *Pl. Soil* 1958, **9,** 353–66.

471 WALKER, T. W.; ADAMS, A. F. R.; ORCHISTON, H. D. Fate of labelled nitrate and ammonium nitrogen when applied to grass and clover grown separately and together. *Soil Sci.* 1956, **81**, 339–51.

472 WALKER, T. W.; EDWARDS, G. H. A.; CAVELL, A. J.; ROSE, T. H. The use of fertilizers on herbage cut for conservation. *J. Br. Grassld Soc.* 1952, **7**, 107–30.

473 WALKER, T. W.; ORCHISTON, H. D.; ADAMS, A. F. R. The nitrogen economy of grass legume associations. *J. Br. Grassld Soc.* 1954, **9**, 249–74.

474 WALSHE, M. J.; CONWAY, A. Hypomagnesaemia in ruminants. *Proc. 8th int. Grassld Congr., Reading, 1960* 1961, pp. 548–53.

475 WATKIN, B. R. The animal factor and levels of nitrogen. *J. Br. Grassld Soc.* 1954, **9**, 35–46.

476 WATSON, E. R.; LAPINS, P. The influence of subterranean clover pastures on soil fertility. 2. The effects of certain management systems. *Aust. J. agric. Res.* 1964, **15**, 885–94.

477 WATSON, E. R.; LAPINS, P. Losses of nitrogen from urine on soils from south-western Australia. *Aust. J. exp. Agric. Anim. Husb.* 1969, **9**, 85–91.

478 WEINMANN, H. Seasonal chemical changes in the roots of some South African highveld grasses. *Jl S. Afr. Bot.* 1940, **6**, 131–45.

479 WEISS, M. G.; MUKERJI, S. K. Effect of planting method and nitrogen fertilization on relative performance of orchardgrass strains. *Agron. J.* 1950, **42**, 555–9.

480 WHEELER, J. L. The effect of sheep excreta and nitrogenous fertilizer on the botanical composition and production of a ley. *J. Br. Grassld Soc.* 1958, **13**, 196–202.

481 WHEELER, J. L. A response to phosphate on an all-grass sward. *J. Br. Grassld Soc.* 1960, **15**, 300–1.

482 WHITEHEAD, D. C. Nutrient minerals in grassland herbage. *Review Series 1/1966 Commonw. Bur. Past. Fld Crops* Farnham Royal: Commonw. Agric. Bur., 1966, pp. 86.

483 WHITEHEAD, D. C. Data on the mineral composition of grassland herbage from the Grassland Research Institute, Hurley, and the Welsh Plant Breeding Station, Aberystwyth. *Tech. Rep. 4 Grassld Res. Inst.* 1966, pp. 55.

484 WHITEHEAD, D. C. Carbon, nitrogen, phosphorus and sulphur in herbage plant roots. *J. Br. Grassld. Soc.* (in press).

485 WHITEHEAD, D. C.; JONES, E. C. Nutrient elements in the herbage of white clover, red clover, lucerne and sainfoin. *J. Sci. Fd Agric.* 1969, **20**, 584-91.

486 WHITT, D. M. The role of bluegrass in the conservation of the soil and its fertility. *Proc. Soil Sci. Soc. Am.* 1941, **6**, 309–11.

487 WIDDOWSON, F. V.; PENNY, A. Anhydrous ammonia. *Rep. Rothamsted exp. Stn 1966* 1967, pp. 41–2.

488 WIDDOWSON, F. V.; PENNY, A.; FLINT, R. C. Experiments on nitrogen fertilizers. *Rep. Rothamsted exp. Stn 1967* 1968, pp. 38–9.

489 WIDDOWSON, F. V.; PENNY, A.; WILLIAMS, R. J. B. An experiment comparing urea-formaldehyde fertilizer with 'nitro-chalk' for Italian ryegrass. *J. agric. Sci., Camb.* 1962, **59**, 263–8.

490 WIDDOWSON, F. V.; PENNY, A.; WILLIAMS, R. J. B. An experiment comparing responses to nitrogen fertilizer of four grass species. *Expl Husb.* 1963, **9**, 28–34.

491 WIDDOWSON, F. V.; PENNY, A.; WILLIAMS, R. J. B. An experiment comparing responses to nitrogen fertilizer of four grass species. Part 2. Residual effects in wheat and barley. *Expl Husb.* 1964, **11**, 22–30.

492 WIDDOWSON, F. V.; PENNY, A.; WILLIAMS, R. J. B. An experiment measuring effects of N, P, and K fertilizers on yield and N, P, and K contents of grass. *J. agric. Sci., Camb.* 1965, **64**, 93–100.

493 WIDDOWSON, F. V.; PENNY, A.; WILLIAMS, R. J. B. Experiments comparing concentrated and dilute NPK fertilizers and four nitrogen fertilizers on a range of crops. *J. agric. Sci., Camb.* 1965, **65**, 45–55.

494 WIDDOWSON, F. V.; PENNY, A.; WILLIAMS, R. J. B. An experiment measuring effects of N, P and K fertilizers on yield and N, P and K contents of grazed grass. *J. agric. Sci., Camb.* 1966, **67**, 121–8.

495 WIDDOWSON, F. V.; SHAW, K. Comparisons of casein and formalized casein with ammonium sulphate, calcium nitrate and urea for Italian ryegrass. *J. agric. Sci., Camb.* 1960, **55**, 53–9.

496 WILLIAMS, C. H.; DONALD, C. M. Changes in organic matter and pH in a podzolic soil as influenced by subterranean clover and superphosphate. *Aust. J. agric. Res.* 1957, **8**, 179–89.

497 WILLIAMS, T. E.; CLEMENT, C. R. Accumulation and availability of nitrogen in soils under leys. *Proc. 1st Gen. Meet. Eur. Grassld Fed. 1965* [n.d.], pp. 39–45.

498 WILMAN, D. The effect of nitrogenous fertilizer on the rate of growth of Italian ryegrass. *J. Br. Grassld Soc.* 1965, **20**, 248–54.

499 WILSON, D. B. Effects of light intensity and clipping on herbage yields. *Can. J. Pl. Sci.* 1962, **42**, 270–5.

500 WILSON, D. B. Interactions in forage yield trials. *Can. J. Pl. Sci.* 1964, **44**, 344–50.

501 WILSON, J. K. The loss of nodules from legume roots and its significance. *Agron. J.* 1942, **34**, 460–71.

502 WILSON, P. W. The biochemistry of symbiotic nitrogen fixation. Madison : Univ. Wisconsin Press, 1940, pp. 302.

503 WILSON, R. K. An attempt to induce hypomagnesaemia in wethers by feeding high levels of urea. *Vet. Rec.* 1963, **75,** 698–9.

504 WIT, C. T. DE; DIJKSHOORN, W.; NOGGLE, J. C. Ionic balance and growth of plants. *Versl. landbouwk. Onderz.* 1963, **69** (15), pp. 69.

505 WOELFEL, C. G.; POULTON, B. R. The nutritive value of timothy hay as affected by nitrogen fertilization. *J. Anim. Sci.* 1960, **19,** 695–9.

506 WOLDENDORP, J. W. The influence of living plants on denitrification. *Meded. LandbHogesch. Wageningen* 1963, **63** (13), 1–100.

507 WOLDENDORP, J. W. Losses of soil nitrogen. *Stikstof* (English edn) 1968, No. 12, 32–46.

508 WOLDENDORP, J. W.; DILZ, K.; KOLENBRANDER, G. J. The fate of fertilizer nitrogen on permanent grassland soils. *Proc. 1st Gen. Meet. Eur. Grassld Fed. 1965* [n.d.], pp. 53–68.

509 WOLTON, K. M. The effect of sheep excreta and fertilizer treatments on the nutrient status of pasture soil. *J. Br. Grassld Soc.* 1955, **10,** 240–53.

510 WOLTON, K. M. Some factors affecting herbage magnesium levels. *Proc. 8th int. Grassld Congr., Reading, 1960* 1961, pp. 544–8.

511 WOLTON, K. M. An investigation into the simulation of nutrient returns by the grazing animal in grassland experimentation. *J. Br. Grassld Soc.* 1963, **18,** 213–19.

512 WOLTON, K. M. Techniques in grassland experimentation. *Bull. Docum. Ass. int. Fabr. Superphos.* 1965, No. 41, 1–13.

513 WOLTON, K. M.; BROCKMAN, J. S.; BROUGH, D. W. T.; SHAW, P. G. The effect of nitrogen, phosphate and potash fertilizers on three grass species. *J. agric. Sci., Camb.* 1968, **70,** 195–202.

514 WRIGHT, M. J.; DAVISON, K. L. Nitrate accumulation in crops and nitrate poisoning in animals. *Adv. Agron.* 1964, **16,** 197–247.

515 YATES, F.; BOYD, D. A. Two decades of surveys of fertilizer practice. *Outl. Agric.* 1965, **4,** 203–10.

516 YOUNG, D. J. B. A study of the influence of nitrogen on the root weight and nodulation of white clover in a mixed sward. *J. Br. Grassld Soc.* 1958, **13,** 106–14.

INDEX